Soft Computing Applications and Techniques in Healthcare

Information Technology, Management and Operations Research Practices
Series Editors: Vijender Kumar Solanki, Sandhya Makkar and Shivani Agarwal

This new book series will encompass theoretical and applied books and will be aimed at researchers, doctoral students, and industry practitioners to help in solving real-world problems. The books will help in the various paradigm of management and operations. The books will discuss the concepts and emerging trends on society and businesses. The focus is to collate the recent advances in the field and take the readers on a journey that begins with understanding the buzz words like employee engagement, employer branding, mathematics, operations, technology and how they can be applied in various aspects. It walks readers through engaging with policy formulation, business management, and sustainable development through technological advances. It will provide a comprehensive discussion on the challenges, limitations, and solutions of everyday problems like how to use operations, management and technology to understand the value based education system, health and global warming, and real time business challenges. The book series will bring together some of the top experts in the field throughout the world who will contribute their knowledge regarding different formulations and models. The aim is to provide the concepts of related technologies and novel findings to an audience that incorporates specialists, researchers, graduate students, designers, experts, and engineers who are occupied with research in technology, operations, and management related issues.

Performance Management
Happiness and Keeping Pace with Technology
Edited by Madhu Arora, Poonam Khurana and Sonam Choiden

Soft Computing Applications and Techniques in Healthcare
Edited by Ashish Mishra, G. Suseendran and Trung-Nghia Phung

Mathematical Modeling and Soft Computing in Epidemiology
Edited by Jyoti Mishra, Ritu Agarwal and Abdon Atangan

For more information about this series, please visit: https://www.crcpress.com/ Information-Technology-Management-and-Operations-Research-Practices/ book-series/CRCITMORP

Soft Computing Applications and Techniques in Healthcare

Edited by

Ashish Mishra, G. Suseendran and Trung-Nghia Phung

CRC Press is an imprint of the
Taylor & Francis Group, an **informa** business

MATLAB® is a trademark of The MathWorks, Inc. and is used with permission. The MathWorks does not warrant the accuracy of the text or exercises in this book. This book's use or discussion of MATLAB® software or related products does not constitute endorsement or sponsorship by The MathWorks of a particular pedagogical approach or particular use of the MATLAB® software.

First edition published 2021
by CRC Press
6000 Broken Sound Parkway NW,
Suite 300, Boca Raton, FL 33487-2742

and by CRC Press
2 Park Square, Milton Park, Abingdon, Oxon, OX14 4RN

© 2021 Taylor & Francis Group, LLC

CRC Press is an imprint of Taylor & Francis Group, LLC

Reasonable efforts have been made to publish reliable data and information, but the author and publisher cannot assume responsibility for the validity of all materials or the consequences of their use. The authors and publishers have attempted to trace the copyright holders of all material reproduced in this publication and apologize to copyright holders if permission to publish in this form has not been obtained. If any copyright material has not been acknowledged please write and let us know so we may rectify in any future reprint.

Except as permitted under U.S. Copyright Law, no part of this book may be reprinted, reproduced, transmitted, or utilized in any form by any electronic, mechanical, or other means, now known or hereafter invented, including photocopying, microfilming, and recording, or in any information storage or retrieval system, without written permission from the publishers.

For permission to photocopy or use material electronically from this work, access www.copyright.com or contact the Copyright Clearance Center, Inc. (CCC), 222 Rosewood Drive, Danvers, MA 01923, 978-750-8400. For works that are not available on CCC please contact mpkbookspermissions@tandf.co.uk

Trademark notice: Product or corporate names may be trademarks or registered trademarks and are used only for identification and explanation without intent to infringe.

Library of Congress Cataloging-in-Publication Data

Names: Mishra, Ashish (Ashish Kumar), editor. | Suseendran, G., editor. | Phung, Trung-Nghia, editor.
Title: Soft computing applications and techniques in healthcare / edited by Ashish Mishra, G. Suseendran and Trung-Nghia Phung.
Other titles: Information technology, management and operations research practices (Series)
Description: First edition. | Boca Raton : CRC Press, 2020. | Series: Information technology, management and operations research practices | Includes bibliographical references and index.
Identifiers: LCCN 2020018646 (print) | LCCN 2020018647 (ebook) | ISBN 9780367423872 (hardback) | ISBN 9781003003496 (ebook)
Subjects: MESH: Medical Informatics Computing | Computing Methodologies
Classification: LCC R855.3 (print) | LCC R855.3 (ebook) | NLM W 26.5 | DDC 610.285—dc23
LC record available at https://lccn.loc.gov/2020018646
LC ebook record available at https://lccn.loc.gov/2020018647

ISBN: 978-0-367-42387-2 (hbk)

ISBN: 978-1-003-00349-6 (ebk)

Typeset in Times LT Std

by Cenveo® Publisher Services

Contents

Preface..vii
Acknowledgements xi
Editors...xiii
Contributors xv

Chapter 1 Analytical Approach to Genetics of Cancer Therapeutics
through Machine Learning 1

Ritu Shukla, Mansi Gyanchandani, Rahul Sahu and Priyank Jain

Chapter 2 A Study on Behaviour of Neural Gas on Images and Artificial
Neural Network in Healthcare 11

Rahul Sahu, Ashish Mishra and G. Suseendran

Chapter 3 A New Approach for Parkinson's Disease Imaging Diagnosis
Using Digitized Spiral Drawing 35

Megha Kamble and Pranshu Patel

Chapter 4 Modelling and Analysis for Cancer Model with Caputo
to Atangana-Baleanu Derivative 57

Ashish Mishra, Jyoti Mishra and Vijay Gupta

Chapter 5 Selection of Hospital Using Integrated Fuzzy AHP
and Fuzzy TOPSIS Method 71

Vikas Shinde and Santosh K. Bharadwaj

Chapter 6 Computation of Threshold Rate for the Spread of HIV
in a Mobile Heterosexual Population and Its Implication
for SIR Model in Healthcare 97

Suresh Rasappan and Regan Murugesan

Chapter 7 Application of Soft Computing Techniques to Heart Sound
Classification: A Review of the Decade 113

Babita Majhi and Aarti Kashyap

Chapter 8 Fuzzy Systems in Medicine and Healthcare: Need,
Challenges and Applications 139

Deepak K. Sharma, Sakshi and Kartik Singhal

v

Chapter 9	Appliance of Machine Learning Algorithms in Prudent Clinical Decision-Making Systems in the Healthcare Industry	163

T. Venkat Narayana Rao and G. Akhila

Chapter 10 Technique of Receiving Data from Medical Devices to Create Electronic Medical Records Database 185

Vu Duy Hai

Chapter 11 Universal Health Database in India: Emergence, Feasibility and Multiplier Effects 215

Arindam Chakrabarty and Uday Sankar Das

Chapter 12 Cluster Analysis of Breast Cancer Data Using Modified BP-RBFN 235

*Viswanathan Sangeetha, Jayavel Preethi,
Raghunathan Krishankumar, Kattur S. Ravichandran
and Ramachandran Manikandan*

Index 257

Preface

It is a pleasure for us to put forth this book, *Soft Computing Applications and Techniques in Healthcare*. In the present era, soft computing approaches play a vital role in solving many different kinds of problems and providing promising solutions. Due to the popularity of soft computing approaches, these approaches have also been applied to healthcare data for effectively diagnosing diseases and obtaining better results in comparison to traditional approaches. A soft computing approach has the ability to adapt itself according to the problem domain. Another aspect is a good balance between exploration and exploitation processes. These aspects make soft computing approaches more powerful, reliable and efficient. The above-mentioned characteristics make the soft computing approaches more suitable and competent for healthcare data. Medical science and engineering have been using various medical systems such as medical imaging devices, medical testing devices and medical information systems. In order to analyse such big data efficiency, image processing, signal processing and data mining play important roles for computer-aided diagnosis and monitoring. In particular, soft computing approaches play a fundamental role in the analysis of such medical data because of the ambiguity of human data. Research is imperative in order to recognise the contemporary issues in such areas and provide insight on better practices and analytical models in various techniques of soft computing in healthcare. This book will enable the reader to appreciate the applications of analytics in soft computing and computational mathematics techniques to identify, categorise and assess the role of different soft computing techniques for diagnosis and prediction of diseases.

- **CHAPTER 1** This chapter proposes that AdaBoost when applied to Random Forest improves the accuracy, while when applied to decision tree, the accuracy was decreased. From this we infer that AdaBoost depends on the algorithm, as decision tree constructed on the dataset using different features, whereas Random Forest is more complicated than decision tree. AdaBoost in combination with Random Forest selects specific features and observation for building up decisions.

- **CHAPTER 2** This chapter discusses many opportunities for ANN, and deep learning and innovators across healthcare (payer and provider) and life sciences (pharma, biotech and medical device manufacturing) are beginning to invest in this area to achieve state-of-the-art accuracy, with many hoping to go beyond human-level performance.

- **CHAPTER 3** This chapter aims to show that three types of digitised spiral drawing tests have a major impact on the classification of Parkinson's disease patients and healthy controls, when four machine learning models are implemented on mathematically processed datasets.

viii   Preface

- **CHAPTER 4** This chapter provides a cancer treatment model through the new fractional derivative which incorporates the contacts connecting well tissue cells, cancer cells and activate resistant classification cells.

- **CHAPTER 5** This chapter presents a general hierarchical model for the selection of the best hospital. Selection under the considerable criteria and subcriteria is examined by using fuzzy AHP and TOPSIS techniques.

- **CHAPTER 6** This chapter focuses on the mathematical model for a mobile heterosexual population. The threshold rate for the transmission of HIV in a mobile heterosexual population has been analysed. HIV transmission of heterosexual female population is considered by means of a simple SIR model.

- **CHAPTER 7** This chapter focuses on classifying heart sound recordings. This is an important area for future research where many new algorithms and techniques will be applied.

- **CHAPTER 8** In this chapter, the detection and treatment of malaria has been studied by many researchers. It not only helps in the timely diagnosis of the disease but also indicates the level of the disease, thereby helping practitioners improve their decision-making process and to restore the health of the patient in the best way possible.

- **CHAPTER 9** This chapter focuses on many new advancements taking place in the field of machine learning that can in turn help provide better healthcare services to the patients.

- **CHAPTER 10** This chapter focuses on the techniques of automatically collecting electronic medical data from various medical devices, implementing, collecting and assessing quality of medical data after it is collected on computers.

- **CHAPTER 11** This chapter focuses on exploring all the angles that can be immensely value added by developing and introducing such comprehensive health database management systems.

Preface

- **CHAPTER 12** This analysis of cancer data to identify cancer survival ability is examined in this chapter.

Dr. Ashish Mishra
Professor, Computer Science and Engineering Department
Gyan Ganga Institute of Technology and Sciences
Jabalpur, Madhya Pradesh, India
Personal Mail ID: ashish.mish2009@gmail.com
Official Mail ID: ashishmishra@ggits.org

Dr. G. Suseendran
Assistant Professor, School of Computing Sciences
Department of Information Technology
Vels Institute of Science, Technology & Advanced Studies (VISTAS)
Pallavaram, Chennai, India
Personal Mail ID: suseendar_1234@yahoo.co.in
Official Mail ID: susee.scs@velsuniv.ac.in

Dr. Trung-Nghia Phung
Associate Professor, Head of Academic Affairs
Thai Nguyen University of Information and Communication Technology
Vietnam
Official Mail ID: ptnghia@ictu.edu.vn

MATLAB® is a registered trademark of The MathWorks, Inc. For product information, please contact:
The MathWorks, Inc.
3 Apple Hill Drive
Natick, MA 01760-2098 USA
Tel: 508-647-7000
Fax: 508-647-7001
E-mail: info@mathworks.com
Web: www.mathworks.com

Acknowledgements

We give sincere thanks to Mrs. Cindy Renee Carelli, Executive Editor, and Ms. Erin Harris, Senior Editorial Assistant, CRC Press, Taylor & Francis Group, for giving us an opportunity to publish this book in the esteemed publishing house, and to "Information Technology, Management and Operations Research Practices" series editors Vijender Kumar Solanki, Sandhya Makkar and Shivani Agarwal for their kind cooperation in the completion of this book. We thank our esteemed authors for having shown confidence in this book and considering it as a platform to showcase and share their original research work.

Editors

Dr. Ashish Mishra is a professor in the Department of Computer Science and Engineering, Gyan Ganga Institute of Technology and Sciences, Jabalpur, Madhya Pradesh, India. He has 16 years of experience in teaching and research and development with a specialisation in Computer Science Engineering. He has a B.E., M.Tech, and MBA. He received his Ph.D. from AISECT University, Bhopal, India. Dr. Mishra has participated in various seminars, paper presentations, research paper reviews, and conferences as co-convener, member of organising committee, member of advisory committee, and member of technical committee. He has contributed in organising the INSPIRE Science Internship Camp. He is a member of the Institute of Electrical and Electronics Engineers and is a life member of Computer Society of India. He has published many research papers in reputed journals and conferences. He also has presented papers at Springer and IEEE conferences. He is also a reviewer and Session Chair of the IEEE international conferences, CSNT-2015, CICN-2016, CICN2017, INDIACom-2019, and ICICC-CONF 2019. His research interests include the Internet of Things, data mining, cloud computing, image processing, and knowledge-based systems. He holds nine patents in Intellectual Property, India. He has authored four books in the areas of data mining, image processing, and LaTex.

Dr. G. Suseendran received his M.Sc. degree in Information Technology and M.Phil. degree from Annamalai University, Tamil Nadu, India. He later received his Ph.D. in Information Technology-Mathematics from Presidency College (Autonomous), University of Madras, Tamil Nadu, India. As an additional qualification, he has obtained a DOEACC 'O' Level AICTE Ministry of Information Technology and Honor Diploma in Computer Programming. He is an assistant professor, Department of Information Technology, School of Computing Sciences, Vels Institute of Science, Technology & Advanced Studies (VISTAS), Chennai, Tamil Nadu, India. He has many years of teaching experience in both undergraduate and postgraduate level. His research interests include ad-hoc networks, the Internet of Things, data mining, cloud computing, image processing, knowledge-based systems, and Web information exploration. He has guided four M.Phil. scholars, and six Ph.D. scholars have been awarded under his supervision. He serves as editor/editorial board member/technical committee member/reviewer of international journals published by Thomson Reuters, SCI, and Elsevier. He served as international committee member towards International Conference conducted in association with IEEE, Springer and Scopus. He has published more than 75 research papers in various international journals such as *Science Citation Index, Springer Book Chapter, Scopus, IEEE Access* and UGC-referred journals. He has presented 35 papers in various international conferences. He has also been awarded six times for his contributions.

Prof. Trung-Nghia Phung received his Engineering degree in Electronics and Telecommunications from Hanoi University of Technology (HUST) in 2002. He completed his Master of Science degree in Telecommunications from the Vietnam

xiii

National University – Hanoi (VNUH) in 2007 and his Ph.D. from Japan Advanced Institute of Science and Technology (JAIST) in 2013. He was Dean of Faculty of Electronics and Telecommunications, Thai Nguyen University of Information and Communication Technology (ICTU). He is now an associate professor and Head of Academic Affairs, ICTU. He has published more than 60 research papers. He was the recipient of the award for the excellent young researcher (Golden Globe Award) from Ministry of Science and Technology of Vietnam in 2008. His main research interest lies in the field of speech, audio, and biomedical signal processing. He serves as a technical committee program member, track chair, session chair, and reviewer of many international conferences and journals. He was a co-Chair of the International Conference on Advances in Information and Communication Technology 2016 (ICTA 2016) and a Session Chair of the 4th International Conference on Information System Design and Intelligent Applications (INDIA 2017).

Contributors

G. Akhila
Sreenidhi Institute of Science and
Technology
Hyderabad, Telangana, India

Santosh Kumar Bharadwaj
Department of Applied Mathematics
Madhav Institute of Technology and
Science
Gwalior, Madhya Pradesh, India

Arindam Chakrabarty
Department of Management
Rajiv Gandhi University (Central
University)
Rono Hills, Doimukh, Arunachal
Pradesh, India

Uday Sankar Das
Department of Management and
Humanities
National Institute of Technology
Arunachal Pradesh
Yupia, Arunachal Pradesh, India

Vijay Gupta
Department of Mathematics
University Institute of Technology
-RGPV
Bhopal, Madhya Pradesh, India

Mansi Gyanchandani
Maulana Azad National Institute of
Technology
Bhopal, Madhya Pradesh, India

Vu Duy Hai
Biomedical Electronics Center (BMEC)
Hanoi University of Science and
Technology
Hanoi, Vietnam

Priyank Jain
Maulana Azad National Institute of
Technology
Bhopal, Madhya Pradesh, India

Megha Kamble
Department of Computer Science and
Engineering
Lakshmi Narain College of Technology
Bhopal, Madhya Pradesh, India

Aarti Kashyap
Department of CSIT
Guru Ghasidas Vishwavidyalaya,
Central University
Bilaspur, Chhattisgarh, India

Raghunathan Krishankumar
School of Computing
SASTRA Deemed University
Thanjavur, Tamil Nadu, India

Babita Majhi
Department of CSIT
Guru Ghasidas Vishwavidyalaya,
Central University
Bilaspur, Chhattisgarh, India

Ramachandran Manikandan
School of Computing
SASTRA Deemed University
Thanjavur, Tamil Nadu, India

Ashish Mishra
Department of Computer Science and
Engineering
Gyan Ganga Institute of Technology
and Sciences
Jabalpur, Madhya Pradesh, India

Jyoti Mishra
Department of Mathematics
Gyan Ganga Institute of Technology
and Sciences
Jabalpur, Madhya Pradesh, India

Regan Murugesan
Department of Mathematics
Vel Tech Rangarajan Dr. Sagunthala
R&D Institute of Science and
Technology
Avadi, Chennai, India

Thota Venkat Narayana Rao
Sreenidhi Institute of Science and
Technology
Hyderabad, Telangana, India

Pranshu Patel
Department of Computer Science and
Engineering
Lakshmi Narain College of Technology
Bhopal, Madhya Pradesh, India

Jayavel Preethi
Department of Computer Science
Anna University Regional Center
Coimbatore, Tamil Nadu, India

Sakshi
Department of Manufacturing Process
and Automation Engineering
Netaji Subhas University of Technology
New Delhi, India

Suresh Rasappan
Department of Mathematics
Vel Tech Rangarajan Dr. Sagunthala
R&D Institute of Science and
Technology
Avadi, Chennai, India

Kattur S. Ravichandran
School of Computing
SASTRA Deemed University
Thanjavur, Tamil Nadu, India

Rahul Sahu
Department of Computer Science and
Engineering
Lakshmi Narain College of Technology
Bhopal, Madhya Pradesh, India

Viswanathan Sangeetha
School of Computing
SASTRA Deemed University
Thanjavur, Tamil Nadu, India

Kartik Singhal
Department of Manufacturing Process
and Automation Engineering
Netaji Subhas University of Technology
New Delhi, India

Deepak K. Sharma
Department of Information Technology
Netaji Subhas University of Technology
New Delhi, India

Ritu Shukla
Maulana Azad National Institute of
Technology
Bhopal, Madhya Pradesh, India

Vikas Shinde
Department of Applied Mathematics
Madhav Institute of Technology and
Science
Gwalior, Madhya Pradesh, India

G. Suseendran
Department of Information Technology
VELS Institute of Science, Technology
and Advanced Studies (VISTAS)
Pallavaram, Chennai, India

1 Analytical Approach to Genetics of Cancer Therapeutics through Machine Learning

Ritu Shukla[1], Mansi Gyanchandani[1],
Rahul Sahu[2] and Priyank Jain[1]
[1]Maulana Azad National Institute of Technology
Bhopal, Madhya Pradesh, India
[2]Department of Computer Science and Engineering
Lakshmi Narain College of Technology
Bhopal, Madhya Pradesh, India

CONTENTS

1.1   Introduction 1
1.2   Literature Review 3
1.3   Data Collection and Processing 4
1.4   Classification and Model Evaluation 4
    1.4.1   K-Nearest Neighbours 4
    1.4.2   Support Vector Machine 4
    1.4.3   Kernels in Support Vector Machine 5
    1.4.4   ADABoost 5
    1.4.5   Random Forest 5
1.5   Logistic Regression 6
    1.5.1   Naive Bayes 7
1.6   Results 7
1.7   Conclusion 8
References 8

1.1   INTRODUCTION

With the rapid increase in cancer, the survival rate has also increased due to the development and advancement of new technologies, surgeries and therapies [1]. These therapies include radiotherapy, chemotherapy and so on. Still, every patient response to treatment is different [2].

Earlier cancer detection techniques include computerised axial tomography (CAT or CT) scans and magnetic resonance imaging (MRI) scans. However, they provided

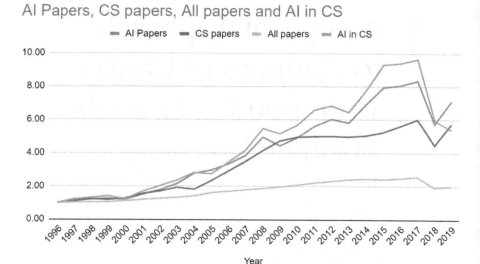

FIGURE 1.1 The growth of annual publications.

much less information about the progression of cancer. Many approaches have been used to find treatment so that the patient can survive. Machine learning (ML) is one of the methods still used for gene expression [3–5] as well as for cancer prediction. Moreover, the accurate prediction of disease can be done by machine learning algorithms.

With the current trend from generalised medicine to personalised medicine, ML techniques can be used for cancer prediction and prognosis. The types of data and ML methods used would increase overall performance. Several studies have been done for early cancer diagnosis [6–11]. The rate at which research is conducted with artificial intelligence (AI) has increased rapidly over the last two decades, as shown in Figures 1.1 and 1.2. Figure 1.2 shows that the major research area is now machine learning.

The accuracy of prediction of cancer has been improved by 15% to 20% over last few years [12].

ML modelling, specifically in AI, has a history in cancer research and practical implementation. A large portion of these works use ML techniques to show the progression of cancer and to recognise information used later in a classification scheme primarily concerning malignant growth, recurrence and survival [13]. Still, ML models suffer from low sensitivity for detecting early-stage cancer and differentiating benign and malignant tumors. Estimation, prediction classification and similar tasks are the major objectives of ML techniques. In particular, ML techniques are commonly used to assign data items into different predefined classes. Misclassification occurs when training and generalisation errors occur. A good classification model should accurately classify all the instances and fit the training set well. This chapter compares various ML algorithms in different aspects to determine which algorithm is best suited for which dataset.

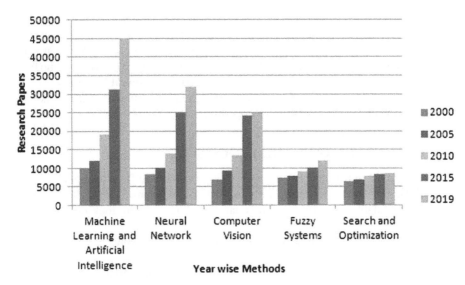

FIGURE 1.2 Scope of machine learning as a major research area has increased over the last few years.

1.2 LITERATURE REVIEW

The use of various ML models in malignant growth research encompasses an tremendous range of applications. Various models dependent on support vector machine (SVM) technology applied to malignant growth forecast issues have been in use for several decades. Different models to predict cancer development and their results have been utilised in several studies. Today data science and bioinformatics commonly use ML-driven models with a wide scope of applications. Studies mostly include classifying, identifying, detecting and distinguishing tumors, predicting cancer and so on.

Breast cancer survival time prediction studies based on ML models occupy a significant part of the contemporary research in this area. There are several studies considering the effect of an ensemble of ML techniques to predict the survival time in breast cancer. Their techniques show better accuracy on their breast cancer dataset compared to previous results [14]. Many papers concern various issues in applying ML algorithms for breast cancer prediction. Researchers experiment on breast cancer datasets [15] using C5 algorithm and achieved 93% accuracy of prediction cancer survivability. Other studies were focused on the comparative analysis of classifiers such as DTs (J48), radial basis function (RBF) neural networks, SVM-RBF kernel and simple classification and regression tree (CART) to find the best classifier. Proper validation is required for the evaluation of the ML algorithms. Performance and accuracy can be achieved by proper validation of ML algorithms. Cross-validation, in particular, is a commonly used method. This method is very suitable for ML-based modelling and is used for training and for testing the datasets [16].

The author [17] received 98.80% and 96.63% accuracies using SVM classification on two different datasets. The ML algorithms Logistic Regression, Naive Bayes, SVM, K-Nearest Neighbours (KNN) comparative study was done by author [13]

4 Soft Computing Applications and Techniques in Healthcare

and was programmed in MATLAB®. [18] used three famous algorithms such as J48, Naive Bayes and RBF to build predictive models on breast cancer prediction and compared their accuracy. The results showed that Naive Bayes predicted well among them with an accuracy of 97.36%. Haifeng Wang and Sang Won Yoon compared Naive Bayes Classifier, SVM, AdaBoost tree and Artificial Neural Networks (ANN) to find a powerful model for breast cancer prediction. They implemented Principal Component Analysis (PCA) for dimensionality reduction [19].

1.3 DATA COLLECTION AND PROCESSING

The dataset used in this chapter was created by Dr. William H. Wolberg, physician at the University of Wisconsin Hospital in Madison, Wisconsin, USA. This dataset is publicly available [20]. To create this dataset the researchers took fluid samples from patients' solid breast masses and performed analysis on Xcyt, which is based on digital scanning of cytological features. The data is heterogeneous and has missing values, so the data had to be normalised.

Data preprocessing was developed using a software module in Python (version 3.7.x) with library scikit-learn for reading and normalising the raw data files.

1.4 CLASSIFICATION AND MODEL EVALUATION

The classifiers used in this chapter are K-Nearest Neighbours, Support Vector Machine, Kernels in Support Vector Machine, AdaBoost and Random Forests. Following is a description of each.

1.4.1 K-Nearest Neighbours

K-Nearest Neighbours (KNN) is a nonparametric lazy learning algorithm, meaning that no training data is used to generalise—that is, training data is required during the testing phase and there is no explicit training phase, or if there is, it is extremely negligible. It takes one instance of the testing data at a time, checks the class of the K-nearest training instances and predicts the class of the testing instance as the one whose occurrences are maximum in the neighbourhood [21, 22].

1.4.2 Support Vector Machine

A widely used algorithm for supervised learning models for identifying pattern and analysing data is Support Vector Machine (SVM). This algorithm is a supervised learning algorithm that is used to identify pattern and analysis of data. Incorporating the algorithm and learning model into this approach enables the problem of classification and regression analysis [23] to be solved. Vapik and his coworker Cortes [24, 25] proposed this method. The main idea behind SVM is to obtain a hyperplane which separates the d-dimensional data perfectly into its classes. An SVM model represents the data as points in space, mapped in such a way that the data of different categories are separated by a clear gap which is as huge as possible. New data is then mapped into that same space and predicted to belong to a category depending on which side of the

Analytical Approach to Genetics of Cancer Therapeutics

separation they occur. [The most basic way to use a SVC is with a linear kernel, which means the decision boundary is a straight line (or hyperplane in higher dimensions).]

1.4.3 KERNELS IN SUPPORT VECTOR MACHINE

It's easy to have a linear hyperplane between the two classes in SVM. But the implementation of SVM is tedious when the classes are nonlinearly separable. For this purpose, important functions called kernels are designed. These functions take low-dimensional input space and convert them into higher dimension space—that is, it transforms nonseparable data into separable data. This approach is called kernel trick. There are multiple categories of kernels available, such as linear, polynomial and RBFs. The linear kernel linearly splits the actual hypothesis into linear functions to output a higher dimensional linear separating hyperplane. In our case C = 10, the classifier is less tolerant to misclassified data points and therefore the decision boundary is more severe.

1.4.4 ADABOOST

AdaBoost was originally called AdaBoost.M1 by the authors of the technique, Freund and Schapire. More recently it may be referred to as discrete AdaBoost because it is used for classification rather than regression. AdaBoost was developed to boost algorithms for binary classification. It is best suited for the performance of decision trees. This is best used for M1 algorithm with weak performance. These models achieve a higher accuracy than its original algorithm.

The most suited and therefore most common algorithm used with AdaBoost are decision trees with one level. Because these trees are so short and only contain one decision for classification, they are often called decision stumps. Each instance in the training dataset is weighted. The initial weight is set to

$$\text{weight}(xi) = 1/n \tag{1.1}$$

where xi is the ith training instance and n is the number of training instances [26–28].

In this chapter, AdaBoost is combined with the decision tree as well as the Random Forest algorithm.

1.4.5 RANDOM FOREST

Random Forest is an ensemble learning method which is used in supervised as well as in unsupervised machine learning. This means it is used in both classification and regression of data. The first algorithm was developed by Tin Kam Ho [29] using the random subspace method, which, in Ho's formulation, is a way to implement the 'stochastic discrimination' approach to classification proposed by Eugene Kleinberg. Further [30] the algorithm was improved by Leo Breiman and Adele Cutler, who trademarked the algorithm as Random Forests [31]. Each tree in this algorithm spits a class prediction, and the maximum votes received by the class become the model prediction. But sometimes classification can be misclassified

6 Soft Computing Applications and Techniques in Healthcare

and performance decreases. In this chapter Random Forest is combined with the AdaBoost algorithm.

The function in decision tree yields a single output – that is, decision – and obtains input as a vector of attribute-value pairs. Usually, the inputs are continuous or discrete; however, for some specific cases described here are discrete for all inputs, which have Boolean values (true or false). They are appropriate for many issues of recognition and classification, where information is complex and the dataset is large [32, 33].

Here, the decision tree classifier selects an attribute as the splitting attribute; it recursively splits the tree using other attributes and organises them in the form of a hierarchical tree structure, where the root nodes represent the class label attribute. The splitting attribute in each level is the one with the maximum information gain.

1.5 LOGISTIC REGRESSION

Logistic regression is a probabilistic linear classifier. The parameter used here is a bias b and weight matrix w. This system enables estimates of categorical outcomes with the assistance of a group of independent variables. [34] The equation for this model is given as

$$y = \text{sgn}(w^T x + b) \tag{1.2}$$

x represents the input given to the system, and y $\mathcal{E}\{-1,1\}$ represents the output class of the system. By using augmented weight matrix θ, equation (1.2) can be written as

$$y = \text{sgn}(\theta^T x) \tag{1.3}$$

Further explanation given below shows that during the period of the training phase of the logistic regression model, augmented weight matrix θ is acquired. Let $\{x_j, y_j\}$ for j = 1, 2... m indicate the training dataset and yj is the target output for training data xj. Initially, weight matrix is initialised as 1, i.e., $\theta = 1$. Weight updates equation is given as

$$\theta_j(n) = \theta_j(n-1) + \alpha \upsilon_j \tag{1.4}$$

where α is the learning rate of the model and υ_j is given as

$$\upsilon_j = \sum_{i=1}^{m} (y^{(i)} - h_\theta(x^{(i)})) x_j^{(i)} \tag{1.5}$$

where m means the number of test samples accessible in preparing dataset, j $\mathcal{E}$ {1,2,...,m} and $h_\theta(x)$ is the logistic function given as

$$h_\theta(x) = \frac{1}{(1 + e^{-\theta T x})} \tag{1.6}$$

Analytical Approach to Genetics of Cancer Therapeutics

1.5.1 NAIVE BAYES

The name Naive Bayes algorithm is derived from the English mathematician Thomas Bayes. This classifier is easier to apply and it is a predictive model [35, 36]. This classification method is based on the theorem of Bayes, which assumes independence between class predictors. Here, we calculate the probability of a particular tuple belonging to a class attribute using the product of the probabilities of the attribute values of the tuple and the probability of the class, using the Naive Bayes theorem. The predicted class of the tuple is the one with the maximum probability.

Let X = {X1, X2, X3 ....Xn} is the sample set, and C1, C2, C3, ...Cm is the class set.

$$P(X|C_i) = \frac{P(X|C_i)P(C_i)}{P(X)} \quad (1.7)$$

The highest probabilities in data samples are calculated for that individual class (1.7).

1.6 RESULTS

If a model's test error rates start to rise even if the training error rates decrease, the model overfitting phenomenon occurs. The training errors of a model can be reduced if the model complexity increases. This chapter compares various results of different machine learning algorithms. Applying different ML classification models, accuracy was checked by the confusion matrix, which basically compares the number of correct predictions to the total number of predictions made. Figure 1.3 shows the accuracy percentage of prediction of cancer. We can infer that Random Forest performance is boosted with combing AdaBoost.

Figure 1.4 shows a comparison of AdaBoost which is applied to Random Forest and Decision Tree. This figure shows that Random Forest when applied with AdaBoost enhances the performance as well as accuracy. Figure 1.5 compares Random Forest with other AdaBoost algorithms. The figure clearly shows that AdaBoost is superior to Decision Tree. In fact, with Decision Tree, the performance has decreased.

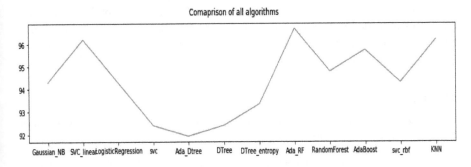

FIGURE 1.3 Percentage comparison of different algorithms.

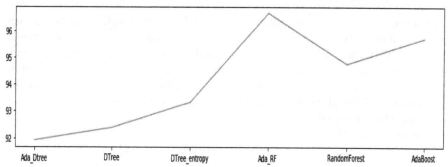

FIGURE 1.4 Comparison of AdaBoost in combination with Random Forest and Decision Tree.

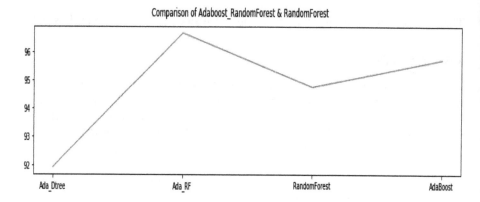

FIGURE 1.5 Comparison of AdaBoost and Random Forest.

1.7 CONCLUSION

Results shows that when applied to Random Forest, AdaBoost improves accuracy, whereas when it was applied to Decision Tree, accuracy decreased. From this we infer that AdaBoost depends on the algorithm, as decision tree is built on the entire dataset using different features whereas Random Forest is more complicated than Decision Tree. Random Forest selects observations/rows and specific features/variables to build multiple decision trees and then averages the results.

REFERENCES

1. Siegel, R. L., Miller, K. D., and Jemal, A. (2016). 'Cancer statistics, 2016'. CA Cancer J. Clin. 66: 7–30. doi: 10.3322/caac.21332.
2. Miller, K. D., Siegel, R. L., Lin, C. C., Mariotto, A. B., Kramer, J. L., Rowland, J. H., et al. (2016). 'Cancer treatment and survivorship statistics, 2016'. CA Cancer J. Clin. 66: 271–289. doi: 10.3322/caac.21349.

3. Mangasarian, Y. J., and Wolberg, W. (2000). 'Breast cancer survival and chemotherapy: a support vector machine analysis'. In D. Z. Du, P. M. Pardalos, and J. Wang (eds.), Discrete Mathematical Problems with Medical Applications, 1999, Vol. 55. DIMACS Center, Piscataway, NJ. pp. 1–10.
4. Abou Tabl, A., Alkhateeb, A., ElMaraghy, W., and Ngom, A. (2017). 'Machine learning model for identifying gene biomarkers for breast cancer treatment survival'. In Proceedings of the 8th ACM International Conference on Bioinformatics, Computational Biology, and Health Informatics, New York, NY: ACM. p. 607. doi: 10.1145/3107411.3108217.
5. Cardoso, F., vant Veer, L. J., Bogaerts, J., Slaets, L., Viale, G., Delaloge, S., et al. (2016). '70-gene signature as an aid to treatment decisions in early-stage breast cancer'. N. Engl. J. Med. 375: 717–729. doi: 10.1056/NEJMoa1602253.
6. Fortunato, O., Boeri, M., Verri, C., Conte, D., Mensah, M., Suatoni, P., et al. (2014). 'Assessment of circulating microRNAs in plasma of lung cancer patients'. Molecules. 19: 3038–3054.
7. Heneghan, H. M., Miller, N., and Kerin, J. M. (2010). 'MiRNAs as biomarkers and therapeutic targets in cancer'. Curr. Opin. Pharmacol. 10: 543–550.
8. Madhavan, D., Cuk, K., Burwinkel, B., and Yang, R. (2013). 'Cancer diagnosis and prognosis decoded by blood-based circulating microRNA signatures'. Front Genet. 4: 116.
9. Zen, K., and Zhang, C. Y. (2012). 'Circulating microRNAs: a novel class of biomarkers to diagnose and monitor human cancers'. Med. Res. Rev. 32: 326–348.
10. Koscielny, S. (2010). 'Why most gene expression signatures of tumors have not been useful in the clinic'. Sci. Transl. Med. 2.
11. Michiels, S., Koscielny, S., and Hill, C. (2005). 'Prediction of cancer outcome with microarrays: a multiple random validation strategy'. Lancet. 365: 488–492.
12. Cruz, J. A., and Wishart, D. S. (2006). 'Applications of machine learning in cancer prediction and prognosis'. Cancer Inform. 2: 59–77.
13. Kourou, K., Exarchos, T. P., Exarchos, K. P., Karamouzis, M. V., and Fotiadis, D. I. (2015). 'Machine learning applications in cancer prognosis and prediction'. Comput. Struct. Biotechnol. J. 13: 8–17.
14. Djebbari, A., Liu, Z., Phan, S., and Famili, F. (2008). 'An ensemble machine learning approach to predict survival in breast cancer'. Int. J. Comput. Biol. Drug Des. 1: 275–294.
15. Liu, Y., Wang, C., and Zhang, L. (2009). 'Decision Tree Based Predictive Models for Breast Cancer Survivability on Imbalanced Data'. In Proceedings of the 3rd International Conference on Bioinformatics and Biomedical Engineering, iCBBE 2009, Beijing, China, 11–13 June 2009, pp. 1–4.
16. Seker, H., Odetayo, M. O., Petrovic, D., Naguib, R. N., Bartoli, C., Alasio, L., Lakshmi, M. S., and Sherbet, G. V. (2002). 'Assessment of nodal involvement and survival analysis in breast cancer patients using image cytometric data: statistical, neural network and fuzzy approaches'. Anticancer Res. 22: 433–438.
17. Osarech, A., and Shadgar, B. (March 2011). 'A computer aided diagnosis system for breast cancer'. Int. J. Comput. Sci. Issues. 8(2).
18. Rana, M., Chandorkar, P., Dsouza, A., and Kazi, N. (April 2015). 'Breast cancer diagnosis and recurrence prediction using machine learning techniques'. Int. J. Res. Eng. Technol. 04(04): 372–376.
19. Wang, H., and Won Yoon, S. Breast Cancer Prediction using Data Mining Method, IEEE Conference paper.
20. Breast Cancer Wisconsin (Diagnostic) Data Set [Online]. http://archive.ics.uci.edu/ml/datasets/breast+cancer+wisconsin+%28diagnostic%29

21. Ilias, B., et al. 'Indoor mobile robot localization using KNN'. In 6th IEEE International Conference on Control System, Computing and Engineering (ICCSCE), IEEE, 2016.
22. Altayeva, A., Suleimenov, Z., and Young, I. C. 'Medical decision making diagnosis system integrating k-means and Naive Bayes algorithms'. In 16th International Conference on Control, Automation and Systems (ICCAS), IEEE, 2016.
23. Fei, Ye. 'Simultaneous Support Vector selection and parameter optimization using Support Vector Machines for sentiment classification'. In 7th IEEE International Conference on Software Engineering and Service Science (ICSESS), IEEE, 2016.
24. Selamat, M. H., Md Rais, H. 'Enhancement on Image Face Recognition Using Hybrid Multiclass SVM (HM-SVM)'. In 3rd International Conference on Computer and Information Sciences (ICCOINS), 2016.
25. Nalavade, K., and Meshram, B. B. (2012). 'Data Classification Using Support Vector Machine'. National Conference on Emerging Trends in Engineering & Technology (VNCET), vol. 2. pp. 181–184.
26. Kégl, B. (20 December 2013). 'The return of AdaBoost.MH: multi-class Hamming trees'. arXiv:1312.6086 [cs.LG].
27. Joglekar, S. (2016). 'AdaBoost—SachinJoglekar's blog'. codesachin.wordpress.com. Retrieved 3 August 2016.
28. Hughes, G. F. (January 1968). 'On the mean accuracy of statistical pattern recognizers'. IEEE T. Inform. Theory. 14(1):55–63. doi:10.1109/TIT.1968.1054102.
29. Ho, T. K. (1995). 'Random Decision Forests'. In Proceedings of the 3rd International Conference on Document Analysis and Recognition, Montreal, QC, 14–16 August 1995, pp. 278–282.
30. Kleinberg, E. (1990). 'Stochastic discrimination'. Ann. Math. Artif. Intel. 1(1–4): 207–239. CiteSeerX 10.1.1.25.6750. doi:10.1007/BF01531079.
31. RANDOM FORESTS Trademark of Health Care Productivity, Inc. - Registration Number 3185828 - Serial Number 78642027: Justia Trademarks.
32. Kumar, D. V., and Jaya Rama Krishniah., V. V. 'An automated framework for stroke and hemorrhage detection using decision tree classifier'. International Conference on Communication and Electronics Systems (ICCES), IEEE, 2016.
33. Zmyslony, M., Wozniak, M., and Jackowski, K. (2012). 'Comparative analysis of classifier fusers'. Int. J. Arti. Intel. App. 3(3): 95–109.
34. Li, Z., Batta, P., and Trajkovic, L. 'Comparison of Machine Learning Algorithms for Detection of Network Intrusions'. International Conference on Systems Man and Cybernetics (SMC) 2018 IEEE. pp. 4248–4253, 2018.
35. Tomar, G. S., Verma, S., and Jha, A. (2006). 'Web page classification using modified naïve Baysian approach'. IEEE TENCON-2006. 1–4, 14–17.
36. Ma, Y., Liang, S., Chen, X., Jia, C. 'The Approach to Detect Abnormal Access Behavior Based on Naive Bayes Algorithm'. In 10th International Conference on Innovative Mobile and Internet Services in Ubiquitous Computing, 2016.

2 A Study on Behaviour of Neural Gas on Images and Artificial Neural Network in Healthcare

Rahul Sahu[1], Ashish Mishra[2] and G. Suseendran[3]
[1]Department of Computer Science and Engineering
Lakshmi Narain College of Technology
Bhopal, Madhya Pradesh, India
[2]Department of Computer Science and Engineering
Gyan Ganga Institute of Technology and Sciences
Jabalpur, Madhya Pradesh, India
[3]Department of Information Technology
VELS Institute of Science, Technology and Advanced
    Studies (VISTAS)
Pallavaram, Chennai, India

CONTENTS

2.1 Introduction to Neural Gas 12
2.2 Neural Network in Healthcare 13
    2.2.1 Current Medical Applications for ANNs 13
        2.2.1.1 Disease Identification and Diagnosis 13
        2.2.1.2 Personalised Medicine 13
        2.2.1.3 Drug Discovery and Manufacturing 14
        2.2.1.4 Predicting and Managing Epidemic Outbreaks 14
2.3 An Introduction to Self-Organising Maps 14
2.4 Related Work on Neural Gas 15
2.5 Proposed Work 16
    2.5.1 Neural Gas Approach 16
        2.5.1.1 Algorithm 16
    2.5.2 Properties of Neural Gas 17
2.6 Experimental Results 17
    2.6.1 Results and Graphs on Different Values of Epochs Parameter 22
        2.6.1.1 Result and Graph for Animal Image [A] 22
        2.6.1.2 Result and Graph for Building Image [B] 23
        2.6.1.3 Result and Graph for Cloud Image [C] 24
        2.6.1.4 Result and Graph for Flower Image [F] 24
        2.6.1.5 Result and Graph on Vehicle Image [V] 24

2.7 Conclusion for Maximum and Minimum Differences 25
    2.7.1 Conclusion Graph for Maximum Differences 25
        2.7.1.1 Conclusion Graph for Maximum Differences for
                Epochs' Values 25
        2.7.1.2 Conclusion Graph for Maximum Differences for Delta
                Values 26
        2.7.1.3 Conclusion Graph for Maximum Differences for
                Iteration (t) Values 26
        2.7.1.4 Conclusion Graph for Maximum Differences for
                Alpha0 Values 27
        2.7.1.5 Conclusion Graph for Maximum Differences for
                Alphaf Values 27
        2.7.1.6 Conclusion Graph for Maximum Differences for
                Lambda0 Values 28
        2.7.1.7 Conclusion Graph for Maximum Differences for
                Lambdaf Values 28
    2.7.2 Conclusion Graph for Minimum Differences 29
        2.7.2.1 Conclusion Graph for Minimum Differences for
                Epochs' Values 29
        2.7.2.2 Conclusion Graph for Minimum Differences for Delta
                Values 29
        2.7.2.3 Conclusion Graph for Minimum Differences for
                Iteration (t) Values 30
        2.7.2.4 Conclusion Graph for Minimum Differences for
                Alpha0 Values 30
        2.7.2.5 Conclusion Graph for Minimum Differences for
                Alphaf Values 31
        2.7.2.6 Conclusion Graph for Minimum Differences for
                Lambda0 Values 31
        2.7.2.7 Conclusion Graph for Minimum Differences for
                Lambdaf Values 32
2.8 Conclusion 32
References 32

2.1 INTRODUCTION TO NEURAL GAS

Neural gas was first introduced in 1991 by T. Martinetz and K. Schulten. Basically, this is a defined procedure for finding the best possible statistics representations build taking place on feature vectors. This procedure was named 'neural gas' because of the tendency of its feature vectors in the adaptation practice to distribute like gas in the data gap. Neural network approaches have been thoroughly studied for many years to obtain greater attainment compared with traditional classical approaches. The neural gas (NG) network algorithm has been shown effective in clustering, vector quantisation [1], speech recognition [2], image processing [3], pattern recognition [4] and topology representation [5] among other applications.

2.2 NEURAL NETWORK IN HEALTHCARE

Artificial neural network (ANN) is an electronic data-processing model that resembles how the humans' biological nervous structure, like the brain, processes information (dendrites, soma, axon, synapses). The heart component of the neural network approach is the original organisation of electronic data-processing structures. Basically, this is the composition of a great number of extremely interrelated processing essentials (neurons) functioning in union to resolve particular tasks. An ANN is developed for a particular function, for instance, pattern recognition or data classification, in the course of a learning practise [6]. Computing procedures, similarly to humans, can use neural networks, to make sense of complex or inaccurate information or data, to mine patterns and to notice trends.

In addition, ANNs offer the following additional benefits:

- Adaptive learning. It can learn strategies to do work based on information specified for training or primary practise.
- Self-organisation: It is can make self-demonstration or demonstration of the knowledge it extracted throughout the learning process.
- Real-time operation: Computations can be processed in parallel, and particularly hardware in the process of modeling and framed can take advantage of this ability.

2.2.1 CURRENT MEDICAL APPLICATIONS FOR ANNs

- Informing clinical diagnosis
- Predicting future disease
- Analysing images

In addition, there are many applications in production, in research as proof of concepts, or being suggested as upcoming potential uses healthcare and life sciences. Several are listed below to offer a perspective on the prospect for ANNs; however, the list is by no means exhaustive:

2.2.1.1 Disease Identification and Diagnosis

- In radiology for disease identification and diagnosis, deep learning systems are trained to detect the presence or absence of disease in medical images and from unstructured text in radiology reports, helping doctors come up with better interpretations [7].

2.2.1.2 Personalised Medicine

- In personalised medicine to treat cancer patients, establishing standards of care and cancer treatment recommendations based upon the latest medical research literature, evidence-based medicine, in combination with the patient diagnosis and medical history.
- Matching patients based upon their diagnosis, medical history and other factors to the optimal clinical trials available locally and nationwide.

2.2.1.3 Drug Discovery and Manufacturing

- Remote monitoring and real-time data access for increased safety, such as monitoring biological and other signals for any sign of harm or death to participants.
- Early-stage drug discovery, for instance, from initial screening of drug compounds to predicted success rate based on biological factors.

2.2.1.4 Predicting and Managing Epidemic Outbreaks

- Monitoring and predicting epidemic outbreaks based on data collected from satellites, historical information on the web, real-time social media updates and other sources.

Primarily, neural networks are collections of plain essentials which are working in parallel. These essentials are motivated by humans' biological nervous patterns. In general, network tasks are deliberated basically by the associations between essentials. The user may give suggestions as well as training neural network to achieve particular functionality by maintaining the parameter values of the associations (weights) among essentials. Usually neural networks are maintained or trained, which is why a unique input value points towards a definite intended outcome

Neural networks are trained and skilled such that they can accomplish difficult functions in diverse areas of functions such as pattern recognition, pattern classification, speech recognition, pattern identification, vision and control systems. Today's neural networks are able to determine problems that are too complicated for traditional computer systems and for humans. Supervised learning is usually adopted, but supplementary networks may be derived with unsupervised learning or with direct modelling procedures. Unsupervised networks are adopted, on behalf of occurrence, to distinguish combinations of data. Some particular linear networks and Hopfield networks are processed directly. In abstract, there is a wide range of designs as well as learning approaches which enhance the options the developer may design. While the research field of neural networks has been used for about 50 years, particularly robust applications have only been developed in the last 15 years, and its research field is rising rapidly. So, this is specifically dissimilar with the branch of control systems or optimisations techniques over which vocabulary, fundamental math and modelling actions are confidently recognised as well as suggested for many years.

2.3 AN INTRODUCTION TO SELF-ORGANISING MAPS

So far, researchers have looked at networks with supervised training approaches, where the goal is that for each input pattern there is one intended output, and the system is capable of making essential outcomes. They are now turning to unsupervised training, where the networks are capable of outlining the individual classifications of training-related information not including any type of help outside of the system. To make this work, it is imagined that class connection is mainly suggested by the input essentials of the sharable most like properties; then the network may be proficient to recognise these properties between the lots of input essentials. Competitive learning is the main type of unsupervised learning, where outcome neurons contend

Behaviour of Neural Gas on Images and Artificial Neural Network

between each other to be triggered, with property that the only one is triggered at similar clock. This perfectly triggered neuron is known as the winner-takes-all neuron or more frequently as the 'winning neuron'. This context may be modelled and executed by importing lateral inhibition connections (also called negative feedback paths) among neurons. Neurons being forced to systematise each other is the best outcome of this procedure. So, this kind of network is called the self-organising map (SOM).

2.4 RELATED WORK ON NEURAL GAS

In T. Martinetz et al. [8], the author proposed a novel approach: a neural network procedure for vector quantisation (modelling of probability density functions) of topographic anatomy relatively and arbitrarily structured manifolds of entered signals which are going to be accessible and fed to a data manifold M, which is a combination of subsets of disparate geometries. In addition, the quantisation of M every neural entity i, i=1,...., N of the system A promotes links explained by Cij $\in \{0, 1\}$ to neural units j among adjoining receptive areas. The resultant connectivity template Cij explains asymptotically the neighbourhood associations among the electronic input data of quantised manifold and defines a diagram which reflects the frequently a priori strange dimensionality and topological composition of electronic data manifold. M. Francesco Camastra et al. [9] suggests a cursive character recogniser, an essential section within several cursive script recognition organisations applicable with a segmentation and recognition method. Character classification is the process of grouping the exercise with the neural gas algorithm (NG) and learning vector quantisation (LVQ). Here NG is adopted for authentication of lowercase and capital versions of a particular letter has coupled into a particular class group. With each and every letter, this process is applied, High probability of finding an optimal quantity of classes exaggerate the truthfulness of an LVQ classifier. A record of 58k typescript is pinned to train and also for testing the designs. The performance obtained lies in a range of uppermost obtainable in the text for the identification of cursive typescript. A development of the truthful classification rate is created by merging the clustering procedure with a classifier. As in this research NG functions better than SOM, which is why it is the preferred clustering procedure. LVQ is handled as a classifier. The most advantageous expression of classes is extracted by figuring out the overlying in the feature room of vectors analogous to superior and minor parts of each character or letter. When overlying has gone to the uppermost level, then both forms can be coupled in a solo class follow-on as an enhancement of the classifier act.

Ajantha S. Atukorale et al. [4] gives details about the NG network. The very first artificial neural network was costly in computing services, as open ordering of the whole distances among synaptic weights as well as the leaning section are necessary. O($NlogN$) is the time complexity with sequential execution. Another proposal estimated for the previously mentioned open ordering, and this time it accepted perfectly. With this proposal, the behaviour of the NG scheme is now quick in sequential execution. In this work, researchers suggested an implicit ranking procedure to boost sequential execution of the actual NG technique. As compared to Kohonen's self-organising feature map (SOFM) method, the steps for converging NG algorithm are fewer, pre-assumption about network structure is not mandatory and global cost function is used recognising its properties.

Researchers also implement the Hong model for an improved classification scheme on inconsistent data. It necessarily required unconstrained calligraphic statistics, as they control conflicting statistics of data within the equal department due to the number of calligraphic techniques. The Hong scheme scientifically takes separation of the input room by presenting data to the upper level, and the training as well as testing specimens were replicated in the upper level, where multiple decision classifications reside for training information. Target classification was adopted by summarising the different classifications schemes produced by second-level network systems. Researchers take schemes with confidence values for the target classification production. This projected scheme was analysed on a calligraphic database which is retrieved in the NIST 19 database.

2.5 PROPOSED WORK

In the present work, research recommends the agenda of the neural gas procedure to distinguish the images. This procedure works on images by finding patterns on images after the neural gas procedure was applied on it. In this research work five subjects of images were taken the subjects of images: (i) animal, (ii) building, (iii) cloud, (iv) flower and (v) vehicle. Different parameters on their different values are used to obtain the results. The parameters are: (i) epochs, (ii) delta, (iii) alpha0 (initial rate), (iv) alphaf (final rate), (v) lambda0 (initial rate), (vi) lambdaf (final rate) and (vii) iteration (t). For achieving the outcome, it requires to obtain and apply the images to identify patterns on images to distinguish the images to each other. The selection of the neural gas technique is like this, as it doesn't require to administer the learning.

2.5.1 NEURAL GAS APPROACH

Neural gas is a biologically adaptive algorithm promoted by Martinetz and Schulten in 1991. It arranges the input signals with relation to how much distance they have maintained. A specific quantity of these are suggested by distance, which is in some ratio; then many adaptation units and strength are dropped according to a fixed state.

2.5.1.1 Algorithm

The main steps for the neural gas procedure are given below:

1. Select and input the image.
2. Input the x-axis and y-axis pixel values to resize the image.
3. a. Input the values of epoch's parameter; the values are 500, 1000, 2000, 3000, 4000, 5000 and 6000 (initial value is 500).
   b. Input the values of the delta parameter; values are 10, 20, 30, 40, 50 and 60 (initial value is 500).
   c. Input the values of the alpha0 parameter; values are 0.1, 0.2, 0.3, 0.4, 0.5 and1.0 (initial value is 0.3).
   d. Input the values of the alphaf parameter; values are 0.01, 0.02, 0.03, 0.04 and 0.05. (initial value is 0.05).
   e. Input the values of the lambda0 parameter; values are 30, 50, 100, 200 and 500 (initial value is 30).

Behaviour of Neural Gas on Images and Artificial Neural Network 17

    f. Input the values of lambdaf parameter; values are 0.01, 0.02, 0.03, 0.04 and 0.05. (initial value is 0.01).

    g. Increase the number of iterations (t); values are 1000, 2000, 4000, 6000, 8000 and 10000 (initial value is 1000).

4. Then the feature values of the image are extracted.

Each of the feature vector (from $k = 0,\dots$, to N-1) is taken as given below:

$$w_{i_k}^{t+1} = w_{i_k}^{t} + \varepsilon \cdot e^{-k/\lambda} \cdot \left(x - w_{i_k}^{t}\right) \tag{2.1}$$

where ε is the adaptation step size and λ is the neighbourhood collections (distance amongst the input signals). ε, λ decrease with rising t. $P(x)$ is a probability distribution of data vectors x, w_i, $i = 1,\dots,$ N are a finite number of feature vectors, t is the time step, i_0 gives the key of the neighbourhood feature vector, i_1 is the indicator of next nearby feature vectors, etc., and i_{N-1} is the indicator of the feature vectors, which are far to x.

5. Repeat the process for the other user.

2.5.2 PROPERTIES OF NEURAL GAS

1. This technique never deletes any node and never creates new nodes.
2. This technique requires smooth tuning of the λ parameters mainly for a good convergence rate and stable structure.
3. Time efficiency is better in this algorithm compared to other approaches.
4. When exceeding convergence ($>t_{max}$), the network node's vector would be representing the distribution that is being represented.
5. This is a network arrangement that represents ANN.
6. With taking the closest feature vector and also all of them with the relation to that, step size getting lower with the continuously raising of the distance in relation, correlated to k-means clustering, a very robust convergence of procedure is targeted.

2.6 EXPERIMENTAL RESULTS

For doing research on different parameters, first, different values of parameters are taken for the animal image and for each different parameter value, neural gas software is executed on MATLAB® version 7.10.0 (R2010a), the absolute values of maximum and minimum difference on different images on different parameters values are taken and graphs are plotted on these values. Following are the observations.

In this research work, implemented algorithm is applied on different subjects of images:

1. Animal (A) (Figure 2.1)
2. Building(B) (Figure 2.2)
3. Cloud (C) (Figure 2.3)
4. Flower (F) (Figure 2.4)
5. Vehicle (V) (Figure 2.5)

FIGURE 2.1 Animal image [A] with a pixel value of 200, 200.

FIGURE 2.2 Building image [B] with a pixel value of 200, 200.

FIGURE 2.3 Cloud image [C] with a pixel value of 200, 200.

FIGURE 2.4 Flower image [F] with a pixel value of 200, 200.

FIGURE 2.5 Vehicle image [V] with a pixel value of 200, 200.

Parameters are chosen to find the pattern of different subject of images (Graphs 2.1–2.7). The parameter and their initial values are:

1. Epochs (e) = 500
2. Delta (d) = 30
3. Iteration (t) = 1000
4. Alpha0 (a0) = 0.3
5. Alphaf (af) = 0.05
6. Lambda0 (l0) = 30
7. Lambdaf (lf) = 0.01

The same procedure is applied for the rest of the subjects of images. Maximum and minimum differences for all parameters and all subjects of images are chosen and represented using graphs. Maximum differences are chosen first and represented using graphs. For the epochs parameter, high value is achieved for the building image and low for the vehicle image. For the delta parameter, high and low values are both achieved for the flower image. For the iteration parameter, high and low

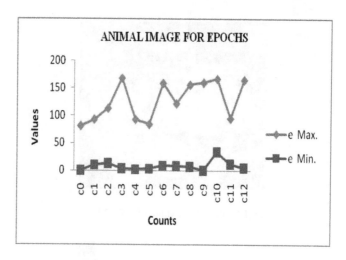

GRAPH 2.1 Animal image for epochs.

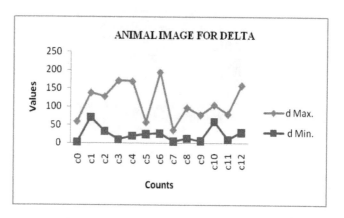

GRAPH 2.2 Animal image for delta.

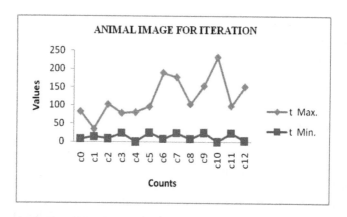

GRAPH 2.3 Animal image for iteration.

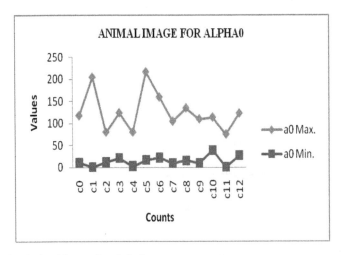

GRAPH 2.4 Animal image for alpha0.

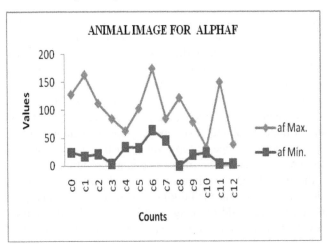

GRAPH 2.5 Animal image for alphaf.

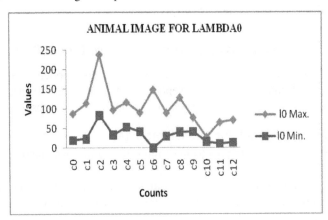

GRAPH 2.6 Animal image for lambda0.

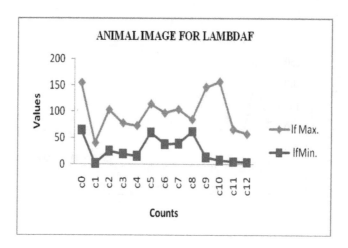

GRAPH 2.7 Animal image for lambdaf.

values are both achieved for the animal image. For the alpha0 parameter, high value is achieved for the building image and low for the cloud image [10]. For the alphaf parameter, high value is achieved for the flower image and low for the animal image. For the lambda0 parameter, high value is achieved for the cloud image and low for the animal image. For the lambdaf parameter, high and low values are both achieved for the building image. After this, minimum difference is chosen and represented using graphs. For the epochs parameter, high value is achieved for the animal image and low value for the animal and cloud images. For the delta parameter, high value is achieved for the flower image and low value for the cloud image and the flower image. For the iteration parameter, high value is achieved for the cloud image and low value for the animal and cloud images. For the alpha0 parameter, high value and low value both are achieved for the building image. For the alphaf parameter, high value is achieved for the building image and low value for the animal and flower images. For the lambda0 parameter, high value is achieved for the cloud image and low value for the animal image. For the lambdaf parameter, high value is achieved for building image and low value for the building and vehicle images. Some of the experimental results with graphs are given below.

2.6.1 Results and Graphs on Different Values of Epochs Parameter

At first, for doing research on the epochs parameter, different epochs values are taken for each subject of images, and for each epoch value, neural gas software is executed on MATLAB® version 7.10.(R2010a) version. Following are the observations.

2.6.1.1 Result and Graph for Animal Image [A]

The other parameters except epochs have the same values as explained previously for all images.

Here the horizontal axis shows the counts, which are the attributes assigned for different colours on a given image after converting it to grayscale, and the

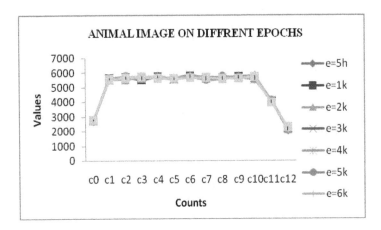

GRAPH 2.8 Animal image on different epochs.

vertical axis shows the values, which gives the total number of same counts on each pixel value (or, say, the same colour on each pixel value) on a given image, The same procedure is applied for different subjects of images and on all different parameters.

Graph 2.8 gives the result on different epoch's values on image A. From the result sheet and graph, the observation is that the maximum number of values occurs between 5000 and 6000 and they are nearly 5500–5800. The minimum values are nearly 2050–2250 at count c12 for all epochs' values.

2.6.1.2 Result and Graph for Building Image [B]

Graph 2.9 gives the result on different epoch's values on image B. From the result sheet and graph, the observation is that the maximum number of values occurs between 5000 and 6000 and they are nearly 5500–5800. The minimum values are nearly 2100–2250 at count c12 for all epochs' values.

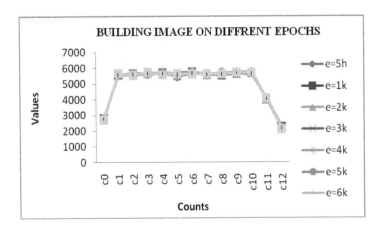

GRAPH 2.9 Building image on different epochs.

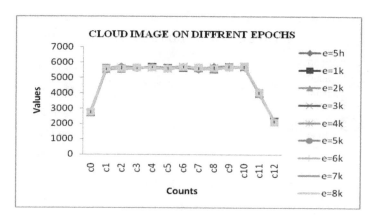

GRAPH 2.10 Cloud image on different epochs.

2.6.1.3 Result and Graph for Cloud Image [C]

Graph 2.10 gives the result on different epochs' values on image C. From the result sheet and graph, the observation is that the maximum number of values occurs between 5000 and 6000 and they are nearly 5500–5800. The minimum values are nearly 2150–2250 at count c12 for all epochs' values.

2.6.1.4 Result and Graph for Flower Image [F]

Graph 2.11 gives the result on different epochs' values on image F. From the result sheet and graph, the observation is that the maximum number of values occurs between 5000 and 6000 and they are nearly 5500–5800. The minimum values are nearly 2100–2250 at count c12 for all epochs' values.

2.6.1.5 Result and Graph on Vehicle Image [V]

Graph 2.12 gives the result on different epochs' values on image V. From the result sheet and graph, the observation is that the maximum number of values occurs

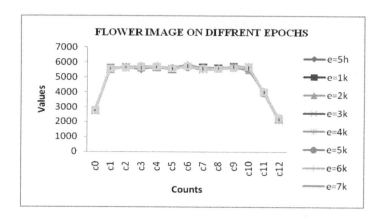

GRAPH 2.11 Flower image on different epochs.

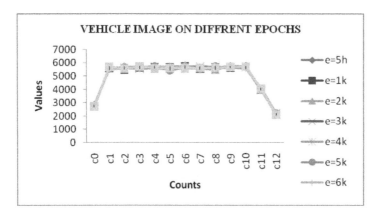

GRAPH 2.12 Vehicle image on different epochs.

between 5000 and 6000 and they are nearly 5500–5800. The minimum values are nearly 2100–2200 at count c12 for all epochs' values.

From all these graphs on different subjects of images on different epochs' values, the conclusion is that the pattern for the graphs are nearly the same. The maximum number of values occurs between 5000 and 6000 and they are nearly 5500–5800.

2.7 CONCLUSION FOR MAXIMUM AND MINIMUM DIFFERENCES

2.7.1 Conclusion Graph for Maximum Differences

2.7.1.1 Conclusion Graph for Maximum Differences for Epochs' Values

Here the maximum difference on epochs' values of images are taken from the tables and plotted on the graph.

From Graph 2.13, the conclusion is that the maximum differences on the epochs of all five subjects of images range from 25–275 nearly. The maximum difference of epochs is of image B at count c5, and the minimum is of V image at count c11.

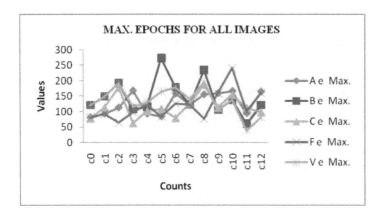

GRAPH 2.13 Maximum epochs for all images.

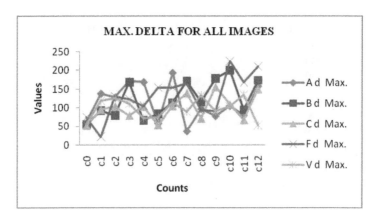

GRAPH 2.14 Maximum delta for all images.

2.7.1.2 Conclusion Graph for Maximum Differences for Delta Values

Here the maximum difference on the delta value of images are taken from the tables and plotted on the graph.

From Graph 2.14, the conclusion is that maximum differences on delta of all five subjects of images range from 25–225 nearly. The maximum difference of delta is of image F at count c10 and the minimum is also of F image at count c1.

2.7.1.3 Conclusion Graph for Maximum Differences for Iteration (t) Values

Here the maximum difference on the iteration (t) value of images are taken from the tables and plotted on the graph.

From Graph 2.15, the conclusion is that the maximum differences on iteration (t) of all five subjects of images range from 25–235 nearly. The maximum difference of iteration (t) is of image A at count c10 and the minimum is also of A image at count c1.

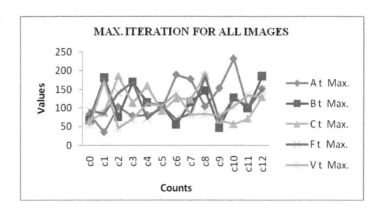

GRAPH 2.15 Maximum iteration for all images.

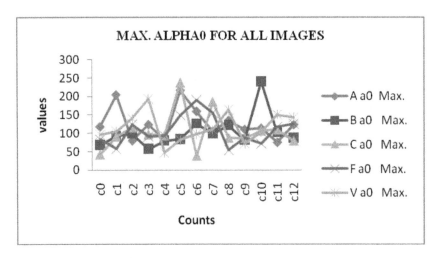

GRAPH 2.16 Maximum alpha0 for all images.

2.7.1.4 Conclusion Graph for Maximum Differences for Alpha0 Values

Here the maximum difference on the alpha0 value of images are taken from the tables and plotted on the graph.

From Graph 2.16, the conclusion is that the maximum differences on alpha0 of all five subjects of images range from 25–250 nearly. The maximum difference of alpha0 is of image B at count c10, and the minimum is of image C at count c6.

2.7.1.5 Conclusion Graph for Maximum Differences for Alphaf Values

Here the maximum difference on the alphaf value of images are taken from the tables and plotted on the graph.

From Graph 2.17, the conclusion is that the maximum differences on alphaf of all five subjects of images range from 30–200 nearly. The maximum difference of alphaf is of image C at count c5, and the minimum is of image A at count c10.

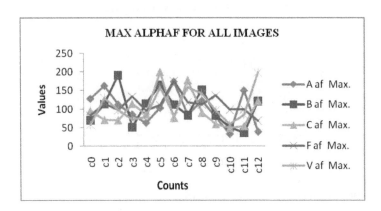

GRAPH 2.17 Maximum alphaf for all images.

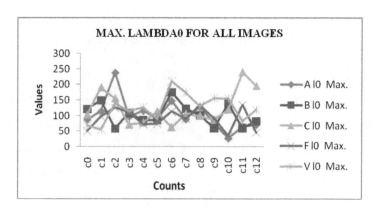

GRAPH 2.18 Maximum lambda0 for all images.

2.7.1.6 Conclusion Graph for Maximum Differences for Lambda0 Values

Here the maximum difference on the lambda0 value of images are taken from the tables and plotted on the graph.

From Graph 2.18, the conclusion is that maximum differences on lambda0 of all five subjects of images range from 25–245 nearly. The maximum difference of lambda0 is of C image at count c11, and the minimum is of image A at count c10.

2.7.1.7 Conclusion Graph for Maximum Differences for Lambdaf Values

Here, the maximum difference on the lambdaf value of images are taken from the tables and plotted on the graph.

From Graph 2.19, the conclusion is that maximum differences on lambdaf of all five subjects of images range from 20–190 nearly. The maximum difference of lambdaf is of image B at count c8, and the minimum is also of image B at count c3.

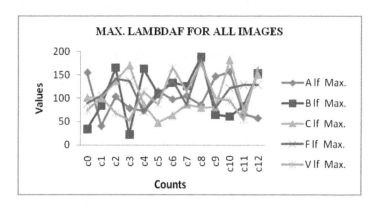

GRAPH 2.19 Maximum lambdaf for all images.

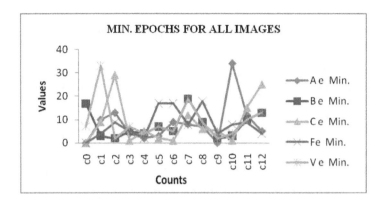

GRAPH 2.20 Minimum epochs for all images.

2.7.2 Conclusion Graph for Minimum Differences

2.7.2.1 Conclusion Graph for Minimum Differences for Epochs' Values

Here the minimum difference on epochs' values of images are taken from the tables and plotted on the graph.

From Graph 2.20, the conclusion is that minimum differences on the epochs of all five subjects of images range from 0–35 nearly. The maximum is of image A at count c10.

2.7.2.2 Conclusion Graph for Minimum Differences for Delta Values

Here the minimum difference on the delta value of images are taken from the tables and plotted on the graph.

From Graph 2.21, the conclusion is that the minimum differences on the epochs of all five subjects of images range from 0–80 nearly. The maximum is of image F at count c11.

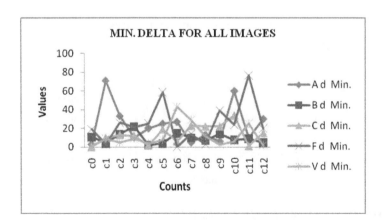

GRAPH 2.21 Minimum delta for all images.

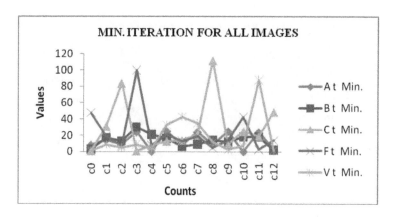

GRAPH 2.22 Minimum iteration for all images.

2.7.2.3 Conclusion Graph for Minimum Differences for Iteration (t) Values

Here the minimum difference on the iteration (t) value of images are taken from the tables and plotted on the graph.

From Graph 2.22, the conclusion is that the minimum differences on epochs of all five subjects of images range from 0–115 nearly. The maximum is of image C at count c8.

2.7.2.4 Conclusion Graph for Minimum Differences for Alpha0 Values

Here the minimum difference on the alpha0 value of images are taken from the tables and plotted on the graph.

From Graph 2.23, the conclusion is the minimum differences on alpha0 of all five subjects of images range from 0–95 nearly. The maximum is of image B at count c10.

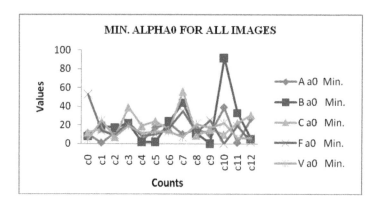

GRAPH 2.23 Minimum alpha0 for all images.

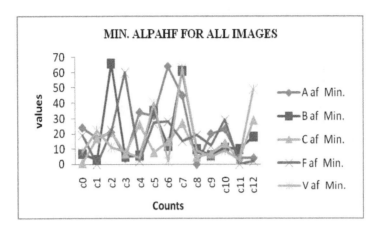

GRAPH 2.24   Minimum alphaf for all images.

2.7.2.5 Conclusion Graph for Minimum Differences for Alphaf Values

Here the minimum difference on alphaf value of images are taken from the tables and plotted the graph.

From Graph 2.24, the conclusion is that the minimum differences on alphaf of all five subjects of images range from 0–70 nearly. The maximum is of image B at count c2.

2.7.2.6 Conclusion Graph for Minimum Differences for Lambda0 Values

Here the minimum difference on the lambda0 value of images are taken from the tables and plotted on the graph. The table and graph are given below.

From Graph 2.25, the conclusion is that minimum differences on lambda0 of all five subjects of images range from 0–85 nearly. The maximum is of image C at count c11.

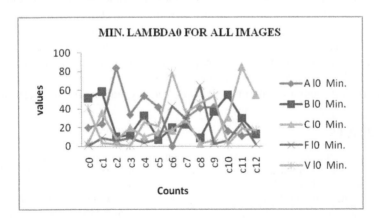

GRAPH 2.25   Minimum lambda0 for all images.

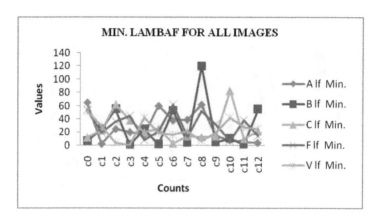

GRAPH 2.26 Minimum lambdaf for all images.

2.7.2.7 Conclusion Graph for Minimum Differences for Lambdaf Values

Here the minimum difference on lambdaf value of images are taken from the tables and plotted on the graph. The table and graph for this are shown below.

From Graph 2.26, the conclusion is that the minimum differences on lambdaf of all five subjects of images range from 0–120 nearly. The maximum is of image B at count c8.

2.8 CONCLUSION

Using the neural gas algorithm with different types of parameters like epochs, delta, iteration, alpha0, alphaf, lambda0 and lambdaf and for all subjects of images like animal, building, cloud, flower and vehicle, the absolute values of maximum and minimum difference on all subjects of images on different parameters values are taken. From this, the conclusion is that when maximum and minimum value of differences are chosen as feature values for images, the parameter epochs, alpha0, alphaf and lambda0 better distinguish the images from each other compared to others and give efficient results. There are many opportunities for ANN and deep learning, and innovators across healthcare (payer and provider) and life sciences (pharma, biotech and medical device manufacturing) are beginning to invest in this area to achieve state-of-the-art accuracy, with many hoping to go beyond human-level performance.

REFERENCES

1. T. Martinetz, S. Berkovich, and K. Schulten. 'Neural-gas network for vector quantization and its application to time-series prediction', IEEE Transactions on Neural Networks, Volume 4, Issue 4, 1993. pp 558–569.
2. F. Curatelli and O. Mayora-Iberra. 'Competitive learning methods for efficient vector quantizations in a speech recognition environment', MICA 2000: Advances in Artificial intelligence: Mexican International Conference on Artificial Intelligence, Acapulco, Mexico, April 2000, proceedings, Springer. pp 109.

3. A. Angelopoulou, A. Psarrou, J. G. Rodriguez, and K. Revett. 'Automatic landmarking of 2D medical shapes using the growing neural gas network', Computer vision for bio-medical image applications: first international workshop, CVBIA 2005, Beijing, China, 21 October, 2005, proceedings, Springer. pp 210.
4. A. S. Atukorale, and P. N. Suganthan. 'Hierarchical overlapped neural gas network with application to pattern classification'. Elsevier. Neurocomputing, Volume 35, 2000. pp 165–176.
5. S. Furao, and O. Hasegawa. 'An incremental network for on-line unsupervised classification and topology learning'. Elsevier. Neural Networks, Volume 19, 2006. pp 90–106.
6. A. K. Qin, and P. N. Suganthan. 'Robust growing neural gas algorithm with application in cluster analysis'. Elsevier. Special Issue, Neural Networks, Volume 17, 2004. pp 1135–1148.
7. Deep Learning Applications in Medical Imaging, 14 September, 2017 by Abder-Rahman Ali: https://www.techemergence.com/deep-learning-applications-in-medical-imaging/
8. T. Martinetz, and K. Schulten. 'A neural-gas network learns topologies'. Artificial Neural Networks, Elsevier Science Publishers B.V., North-Holland, 1991. pp 397–402.
9. F. Camastra, and A. Vinciarelli. 'Combining neural gas and learning vector quantization for cursive character recognition'. Elsevier. Neurocomputing, Volume 51, 2003. pp 147–159.
10. A. Mishra. 'An Authentication Mechanism Based on Client-Server Architecture for Accessing Cloud Computing'. International Journal of Emerging Technology and Advanced Engineering (ISSN 2250-2459), Volume 2, Issue 7, July 2012. pp 95–99.

3 A New Approach for Parkinson's Disease Imaging Diagnosis Using Digitized Spiral Drawing

Megha Kamble and Pranshu Patel
Department of Computer Science and Engineering
Lakshmi Narain College of Technology
Bhopal, Madhya Pradesh, India

CONTENTS

3.1 Introduction 35
    3.1.1 Digital Spiral Basics 36
    3.1.2 Text Organisation 37
3.2 Existing Methods and Experiment Performed 37
    3.2.1 Related Work from the References 37
    3.2.2 Performed Work 41
        3.2.2.1 Data Acquisition and Collection 41
        3.2.2.2 Feature Engineering 42
        3.2.2.3 Model Selection and Parameter Tuning 50
        3.2.2.4 Experimental Setup 52
3.3 Results 52
    3.3.1 Metrics 52
3.4 Discussion 54
3.5 Conclusion 55
References 55

3.1 INTRODUCTION

Spiral drawing is a practised and composite coordinated motor activity. Therefore, it is treated as sensitive motor assessment. Motor rating scale and its subscale Unified Parkinson's disease rating scale (UPDRS-III) is the popularly used and reliable measuring scale in Parkinson's disease (PD). PD affects various functions of the body such as speech, handwriting, walking and coordination movements, and all these are considered part of motor functions. All quantitative indices of motor decline and non-motor biomarkers have been proposed to check the degree of severity, as PD is treated as a motor disorder due to the neurodegeneration process. The diagnosis

36 Soft Computing Applications and Techniques in Healthcare

and monitoring of PD is an expensive and tiresome process for two reasons. First, it is inconvenient for caretakers to take the patient to the clinic and perform physical examinations and, second, diagnosis based on physical observation requires trained medical experts. A clinical invasive technique is available only at the early stage, but it incurs risk with limited resources available in underdeveloped parts of the world, and if early diagnosis is done, only then it is useful.

Hence, previously, PD was assessed by traditional noninvasive methods such as handwriting tests and spiral drawing pen and paper tests. Collecting, preserving the paper and analysing these drawings is purely based on human expertise, which is a time-intensive, less accurate, biased method. With the advancement of IT, it is easier to collect the drawing samples using digitizing (digital) tablets, which are already used in biomedical research. The use of graphics tablets enables researchers to develop various image analysis and processing tools for obtaining different kinds of information that cannot be easily performed with pen and paper drawings.

3.1.1 Digital Spiral Basics

The graphics tablet is equipped with specially designed software that records spiral drawings and generates data files (say, CSV format) with coordination parameters of pen grip angle, pressure, time consumed or current timestamp and x, y, z coordinate values of every stroke stored. This can be a self-administered and noninvasive application that enables PD patients or caretakers to collect data at home and transmit it over the internet to dedicated applications or servers for processing. Here we propose the detection method of the presence of PD by image processing. There are three types of spiral drawing tasks suggested clinically [1]. The data files and drawings of static spiral and dynamic spiral drawings are available on kaggle. With the help of the mathematical model for kinematics, additional coordination parameters related to tremor are taken out. Classification and prediction algorithms are applied on this modified dataset and drawings for separating healthy control and PD patients.

- Here the image analysis is done using the HOG technique. The histogram of oriented gradients (HOG) is a feature descriptor applied in the domains and subdomains of computer vision, image analysis, graphics processing and manipulation for the purpose of uncovering the objects. In this procedure, the amount of gradient orientation is computed particularly in confined small areas of an image. HOG partitions an image into minute square-shaped areas or cells and figures a histogram of oriented gradients in each cell. To return a prominent feature descriptor for each cell, it normalises the result using a blockwise pattern in each small area.
- Identification of three main motor symptoms is the primary objective:
  - Tremor (cannot be directly identified from the features given by the digitized graphics tablet)
  - Muscular rigidity (cannot be directly identified from the features given by the digitized graphics tablet)
  - Slowness and weakness of movements (can be identified by the time taken to complete the task, average pressure and grip angle)

Parkinson's Disease Imaging Diagnosis Using Digitized Spiral Drawing

- Highlights
  - Spiral hand movements are used to aid in the diagnosis of PD.
  - Exerted pressure on the surface and grip angle through spiral hand movement are mapped to kinematic features and converted into an extensive dataset contributing to the diagnosis of motor disorder caused by PD.
  - The dataset is a combination of three tests: static, dynamic and stability drawing. So, the model is flexible for any type of test. The dataset is in the form of CSV files and PNG file format for drawings.
  - Feature extraction is done with the unique feature extraction mechanism of HOG, as well as introduced unique features related to radial velocity, moderately supporting to diagnosis.
  - Exclusive methods based on image analysis of simple spiral drawing can further be extended for diagnosis of other neurodegenerative diseases.
  - Our contribution is applying the most correlated unique features extraction technique on flexible drawing datasets combining all tests for more accurate diagnosis.

A spiral drawing database containing samples from two datasets—15 PD and 15 healthy subjects and another set composed of 25 PD and 15 healthy subjects—is presented, and results are demonstrated for the evaluation of measures such as accuracy, precision, recall and F1 score for discrimination patients and control subjects.

3.1.2 TEXT ORGANISATION

The remainder of this chapter is organised as follows. Section 3.2 gives a description of existing references and methods already pointed out in the literature, with a proposed work framework. Section 3.3 provides an extensive list of features mapped from existing datasets, experimental results and discussions. Section 3.4 discusses limitations of our system and presents conclusions with future scope.

3.2 EXISTING METHODS AND EXPERIMENT PERFORMED

3.2.1 RELATED WORK FROM THE REFERENCES

This section provides details of the methodology in the primary domains of machine learning and neural network used, along with information, datasets, sets of images and previous efforts for PD diagnosis with corresponding references.

The significant work is done in analysing handwriting samples and speech samples for evaluation of tremor and degree of severity of PD, with the conventional techniques of neural network and regression. To improve the correctness of the diagnosis, the work can be extended for scientific diagnosis tool of PD, that is, spiral drawing. Machine learning models and ensemble learning models prove better in predictions and diagnosis supported by feature engineering for revealing more details of symptoms.

The scientific method of spiral drawing comprises three types: [1] Stability Test on a Certain Point (STCP), [2] spiral test static and [3] dynamic test.

Peter Drotar et al. suggested analysis of PD patient handwriting by applying features of selection algorithm and machine learning SVM [2, 3]. This is one of the early works in identifying the importance of in-air/on-surface hand movements in diagnosis of motor disorder of neurodegenerative diseases. The results of the study by Drotar et al. demonstrated that these movements have a major impact during the assessment of handwriting and illustrated 85.61% prediction accuracy in [2]. The work presented PaHaW handwriting's database generated by allowing PD patients to perform eight different handwriting tasks including the Archimedean spiral. The work [3] investigated new pressure features and stroke features related to dynamics of handwriting in extensive manner. Then, three different classifiers, KNN, AdaBoost ensemble and SVM, were applied to demonstrate 81% classification accuracy. However, other demographic features such as age, education, job and birthplace, as well as disease duration to illustrate more practical and effective diagnosis of PD patient must support the handwriting test. In lieu of this handwriting experiment, Archimedean spiral-based scientific method of static and dynamic spiral can be assessed as more relevant with the help of powerful ML methods.

Donalto Impedovo et al. also presented handwriting as a powerful marker to develop a PD diagnostic tool [1, 4]. The authors have implemented the ML classification framework on the same PaHaW dataset for good specificity performance measures. The work not only illustrates binary discrimination of the patients but the work is extended to group the patients based on degree of severity of the illness to further support early cure to diseased persons. Though work has provided an extensive dataset, and grouping based on illness for proper classification, scientific tools of spirals can be extended for simple performance.

Poonam Zham et al. presented and investigated kinematic features of micrographia associated with PD, during repetitive writing tasks [5]. The experimental setup demonstrated samples of the letter 'e' similar to spiral, and the average size of first five and last five letters were compared. These handwriting samples are assessed by Wilcoxon signed-rank test and Kruskal-Wallis test.

Omer Eskidere et al. described applications of machine learning frameworks SVM, LSSVM, Multilayer Perception Neural Network (MLPNN) and general regression NN for remote tracking of PD progression [6]. The work is implemented on telemonitoring dataset composed of early-stage PD patients' voice recordings in 26 attributes. Performance measures mean absolute error (MAE), MSE (mean square error) and correlation coefficients are applied to evaluate the prediction model to conclude that LS-SVM is superior in mapping UPDRS with vocal features. It works on classical and nonclassical, nonlinear voice features. Feature extraction and normalisation of voice features is a complex process, requiring a wider range for normalisation, and accuracy level depends on the signal processing algorithm. This is the traditional mechanism to diagnose PD, and in early stages the symptoms may not be evident in the patient. So UPDRS mapping with only voice features may not lead to accurate diagnosis. The two common drawbacks with the ML algorithms presented in this chapter are a greater computational burden and proneness to over fitting.

Mohammed Erdem Isenkula et al. suggested improved spiral test using digitizing tablet over traditional static spiral test with pen and paper [7]. New Dynamic Spiral Test is suggested by the author. Dynamic spiral test was performed only with a tablet, and the paper has concluded that digital samples generated by computer system

Parkinson's Disease Imaging Diagnosis Using Digitized Spiral Drawing

DST drawings combined with SST drawings can be used to build a generic PD tele-monitoring, diagnosis and controlling system. The tablet specifications provided in the paper are resolution of 1000 pts/cm, accuracy 0.025 cm and 256 levels of measurable pressure connected to computer via USB using proprietary software to process the spirals. Ideal spiral and PD patient spirals are transformed into radius angle transformation that is mathematically equivalent to spiral express linearity in terms of r, θ (theta) polar coordinates. The paper is a significant step towards a noninvasive remote monitoring tool for early-stage diagnosis of PD, the most dangerous neurodegenerative disease. However, this chapter has suggested a framework for combining SST and DST drawings for image processing for PD diagnosis. The experimental setup is not provided to move ahead with diagnosis work. The proposed work in this chapter is providing a solution to effectively use SST and DST drawings of healthy and diseased persons for discrimination of these subjects for clinical assessment at early stage.

Marta San Luciano et al. suggested the use of spiral drawing for computer analysis of PD as digitized spirals correlate with motor scores [8]. Generated/derived indices correlated with overall spiral execution are kinematic irregularity in terms of second-order smoothness, first-order zero crossing, shape, severity, tightness, spiral mean speed and width variability.

Cross-validation performed with linear mixed model to examine validity of combined indices, discriminative validity performance measure was opted with subparameters sensitivity and specificity to conclude early PD diagnosis in the samples and proved spiral analysis accuracy for discrimination of subjects with PD and early PD from controls.

Samayeh Aghnnavesi et al. presented their work aiming to confirm and explore clinometric characteristics of the proposed method applying the entropy mechanism. It measures the disorder caused due to PD associated with upper-limb temporal disorder while depicting spirals using digitized tablets [9]. The work demonstrated calculations of temporal irregularity score (TIS) and differences in mean TIS between diseased and fit subjects. The significant difference of TIS between healthy and advanced PD subjects demonstrate reliability of measures of TIS for relatively small dataset.

F. Miralles et al. developed a fresh quantitative analysis tool of spiral drawing drawn over a print template [10]. Cross-correlation coefficient, mean and standard deviation of actual drawing and template drawing and Fourier transformation are the methods applied for analysis. Results of tremor analysis with scanned image drawings is illustrated with receiving operating characteristic (ROC) curve. This is one of the early experiments done for evaluation of patients' movement disorders. However, a major disadvantage of Fourier transformation is that it is difficult to ensure the subtle changes in the frequency over time when applied to scanned image. Therefore, it cannot explore more dynamics of the movement. In addition, these scanned images require preprocessing for accurate results and analysis.

Saunders-Pullman et al. have validated spiral drawing to propose a measurement scale for motor dysfunction in PD [11]. Several spiral indices such as smoothness, crossing, degree of severity and mean spiral speed are calculated to compare them with clinical scale motor UPDRS, and results demonstrated that there is significant correlation of spiral drawings with assessment of PD and thus given a detailed analysis for PD diagnosis with non-invasive spiral drawing test. This is pen and paper test, and statistical measures were applied to it. The drawback is the presence of outliers, and so to have more accurate results, digital spirals and ML methods are found to be more suitable for PD diagnosis.

Zham et al. have presented guided Archimedean spiral drawing for PD patients and recorded data corresponding to n samples required to draw it [12]. Along with conventional kinematic features, Zham et al. introduced amalgamated key of pace and stress (CISP) to evaluate severity of the PD. The results demonstrated that this is a novel feature with strong correlation with motor ability deterioration due to PD. The limitations of the work are dependence on specially designed spiral tool, real-time results can demonstrate the current situation, progressive information cannot be derived and few samples are available which are to be extended for analysis that is more extensive.

The objectives of Thomas et al. are to present an outline of the methods and technologies used nowadays for studying patterns of handwriting for movement disorder diseases and focusing all the possible parameters related to handwriting that may provide research direction for accurate diagnosis and cure for neural degenerative diseases such as Parkinson's, cerebral palsy and so on [13]. The study by Thomas et al. is a review paper stating different feature extraction mechanisms for handwriting and exploring the future possibility about signature analysis, visual abilities and attention abilities based on handwriting analysis. The kinematic features and datasets presented in the literature are mostly based on demographic features and so might affect evaluation of PD.

Drotar et al. collected handwriting samples from 37 medicated PD patients and 38 age- and sex-matched controls by performing seven handwriting tasks [14]. Conventional kinematic and spatial-temporal handwriting measures and some computed novel handwriting measures based on entropy, signal energy and empirical mode decomposition of the handwriting signals are generated. Selected features when classified with SVM classifier provided 88% accuracy with highest sensitivity and specificity. However, the medicated PD patients' data need to be reevaluated by some reliability tests for more accurate results and extensive feature extraction only contributed to complexity of the model without significantly improving the performance.

João et al. expressed the critical effects of Parkinson's disease on many people around the globe and as a result formulated a strategy to help diagnosing the disease [15]. João et al. focused on extracting features from handwriting exams and also presented its important contribution with results as contribution to the diagnosis of these kinds of neuro diseases. João et al. proposed the likeness measure between the assessment template and the handwritten sketch of the patient for the template. This similarity measure was determined using the structural co-occurrence matrix mathematical model. The said approach evaluated a number of traces and case studies of the patients. Each of these samples were put together with the ML methodology for classification SVM and naive Bayes. João et al. conclude with a promising note about handwriting exam template tools as best contributor for the diagnosis of Parkinson's disease.

The proposed work avoids the handwriting samples for PD diagnosis as demographics features; literacy of the person is important in the case of handwriting, and so it cannot be considered as an independent tool for PD diagnosis. The proposed work suggests scientific tool of SST, DST and STCP to identify diseased persons. The literature has provided handwriting analysis and Archimedean guided spiral analysis. However, literature has not provided diagnosis using digitized recordings of scientific biomarker—in other words, spirals for PD patients. To make the work extensive, additional kinematic features

Parkinson's Disease Imaging Diagnosis Using Digitized Spiral Drawing 41

inspired by [2, 3] are calculated using mathematical models for kinematics and the latest technique of computer vision histogram of oriented gradients. This feature engineering has converted the spiral drawing dataset into a precise dataset that can be applied in the future by researchers for effective early diagnosis of PD.

3.2.2 Performed Work

3.2.2.1 Data Acquisition and Collection

Spiral drawing dataset I consists of drawing samples and a digital record of 25 PDP (Parkinson's disease patients) and 15 healthy individuals is downloaded from a UCI machine learning repository. The dataset is provided by the authors [9] from Istanbul University, at the resource https://archive.ics.uci.edu (ML datasets for Digitized Graphics Tablet-generated Parkinson Disease Spiral Drawings). Spiral drawing dataset II consists of drawing samples and a digital record of 15 PDP (Parkinson's disease patients) and 15 healthy individuals is downloaded from kaggle.

From all the age-matching persons, three types of drawings are drawn on digitized tablets. These three types of spiral drawing tests are based on scientific base for motor disorder assessment and are as follows:

1. Static Spiral Test (SST): Three wounds of an Archimedean spiral appear on the graphics tablet and persons are asked to retrace them with a digital pen.
2. Dynamic Spiral Test (DST): In this type of test, unlike a static test, the spiral appears and disappears in certain time intervals. The patient has to retrace the blinking spiral by keeping the pattern in the mind. This determines the pattern-memorising skills of the patient and pause time taken by the patient while retracing the same pattern.
3. Stability Test on Certain Point (STCP) or Circular Motion Test: There is a red point in the middle of the screen and patients are to draw the circle without touching the screen. This determines the hand stability of the patient.

During the test, the specialised software designed using API and C# records the information of the drawing. The drawings—PNG images of SST and DST and text files (CSV format)—are appended to the dataset. The same dataset of text files is used in the proposed work.

Handwriting dataset data files and drawings were collected using a Wacom Cintiq 12WX graphics table [7]. It is a digital graphics tablet with a rolled-in LCD monitor. It interacts with digitized pens to display a PC monitor on screen. The device sampling was event-based. A sensor incident was marked every time the sensor values x and y were changed.

The digital readings recorded are x, y coordinates, triaxial recording of pressure axes, pressure (unit less), stylus grip angle, timestamp and test id (SST-0, DST-1, STCP-2). The geometric location(x, y) of the pen is at definite time stamps. At the same time, the exerted pressure over the drawing surface, pen tendency, and the in-air movement performed by the pen are the kinematic representations (see Figure 3.1).

Time consumed to complete drawing can be calculated using individual stroke timestamp recorded in the test file w.r.t for each spiral drawing. In the literature,

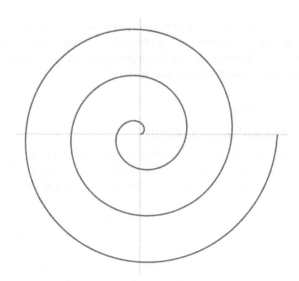

FIGURE 3.1 Spiral drawing test template.

SST and DST images are analysed based on speed and acceleration instantaneous acceleration as velocity changes dramatically.

3.2.2.2 Feature Engineering

Samples stored in the text file and samples required to complete the drawings are longer in the case of SST as compared to DST (see Figures 3.2 and 3.3). This lacks kinematic features, which are based on three factors: instantaneous velocity, acceleration and fluency. So, other features, inspired by [2, 3], are generated using mathematical models and are shown in the following list.

3.2.2.2.1 Kinematic Features

- Duration of static and dynamic spiral (calculated from timestamp provided in CSV file)
- A stroke is a single attached and enduring feature trait of the spiral-drawing pattern.
    - The number of strokes is calculated by counting the number of times on-surface pressure is changing during the whole spiral drawing.
    - The strokes that are surface-based are matched with the outline left on the tablet surface.
    - There are unreal or abstract in-air traces that are stating the pauses and gaps between the letters and also uncertainty in drawing ability or indecisive behaviour while drawing.

Parkinson's Disease Imaging Diagnosis Using Digitized Spiral Drawing

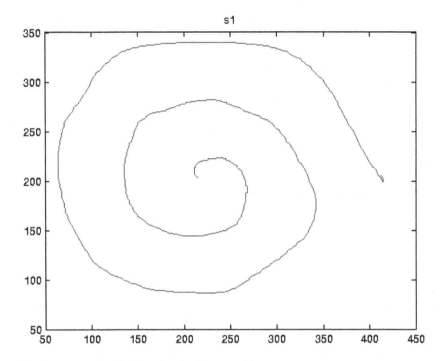

FIGURE 3.2 Static spiral drawing (patient sample).

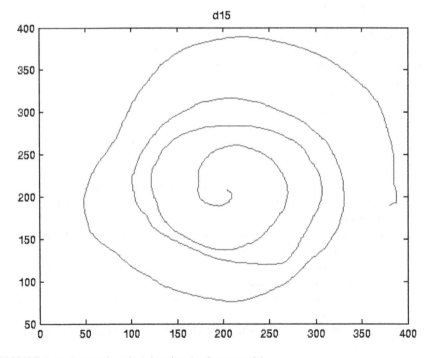

FIGURE 3.3 Dynamic spiral drawing (patient sample).

- Stroke speed: Using python numpy library function linalg.norm, total distance covered by spiral drawing is calculated with the help of x, y and z coordinates available in the CSV file and divided by total time duration, also available in the timestamp field in the CSV file.
- Stroke velocity, acceleration, jerk.

Velocity is a vector quantity and it is averaged as the sum total displacement divided by the time duration required to do so. Displacement can be defined as the straight-line distance between two following sampled coordinates.

The velocity at each spiral point is calculated in this mathematical model as the square root of the sum of squares of successive x, y point coordinates over the time difference recorded for these points in any unit.

$$speed = \sqrt{(x_{i+1} - x_i)^2 + (y_{i+1} - y_i)^2} \div (t_{i+1} - t_i) \qquad (3.1)$$

Based on this displacement, quite a few kinematic factors have been computed in this proposed work by referring to [3], and they are listed as quantity of total strokes; displacement in horizontal and vertical direction, jerk and velocity; acceleration for writing ability; and number of changes of velocity/acceleration (NCV/NCA), NCA and NCV relative to writing duration.

Average acceleration is the change in velocity divided by an elapsed time.

Jerk is the change in the acceleration over time.

The horizontal and vertical components are:

- Magnitudes of horizontal velocity
- Magnitude of vertical velocity
- Magnitude of horizontal acceleration
- Magnitude of vertical acceleration
- Magnitude of horizontal jerk
- Magnitude of vertical jerk

  - Magnitude of horizontal velocity [equation (3.1)] is calculated by mean displacement in horizontal direction (change of x coordinate) over the total time duration.
  - Magnitude of vertical velocity is calculated by mean displacement in vertical direction (change of y coordinate) over the total time duration.
  - Magnitude of horizontal acceleration is mean change of velocity in horizontal direction over the time.
  - Magnitude of vertical acceleration is mean change of velocity in vertical direction over the time.
  - Magnitude of horizontal jerk is mean change of horizontal acceleration over the time.
  - Magnitude of vertical jerk is mean change of vertical acceleration over the time.

Parkinson's Disease Imaging Diagnosis Using Digitized Spiral Drawing 45

- Overall new dataset 38consisted of 3 features (given by original dataset—pressure, grip angle, time) and 16 new features are engineered. Thirty-eight columns (19*2) for two spiral tests are added.

Our contribution in this feature engineering is the addition of two kinematic features, radial velocity and change in radial velocity with respect to time, which are helpful in the diagnosis of hand motion and in turn hand tremor of the PD patient. The two features introduced are as follows:

1. Radial velocity: The radial velocity of an object with respect to a given point is the rate of change of the distance between the object and the point. A spiral drawn by a patient or healthy control is analysed for this feature with reference to origin. Radial velocity is calculated by dividing distance between two consecutive points (displacement) by signed angle between the ray ending at the origin and passing through the point (x, y) on spiral. Python numpy library function atan2 method returns a numeric value in radians representing the angle between the ray passing through (x, y) and positive x axis.
2. Normalised radial velocity: Radial velocity with time is the change of radial velocity between any two consecutive points on spiral over the time elapsed.

Radial velocity and radial velocity with time are extended to mean value and standard deviation values.

3.2.2.2.2 Spatio-Temporal Features

- Number of changes in the velocity: This is a normalisation feature, storing all those instances where velocity change is significant (i.e., more than zero), for static and dynamic spiral drawing tests.
- Number of changes in the acceleration: This is a normalisation feature, storing all those instances where acceleration change is significant (i.e., more than zero), for static and dynamic spiral drawing test.
- On-surface time is applicable for static and dynamic spiral test, whereas in-air time is applicable for STCP (stability test).
- STCP test component in-air time: Pressure exerted with pen (value less than 600 represents pen is in air and no surface pressure exerted), so, data values less than 600 will be counted as in-air time.
- On-surface time for static spiral test: Pressure data values more than 600 are counted from CSV file static spiral test drawings and normalised.
- On-surface time for dynamic spiral test: Pressure data values more than 600 are counted from CSV file dynamic spiral test drawings and normalised.

Table 3.1 shows the correlation factor of the number of features introduced in the literature [2, 3] playing the role in diagnosis of PD, in the proposed models.

TABLE 3.1
Correlation Factors

Features	Correlation Factors
Number of strokes of static spiral test	0.197
Number of changes in velocity of dynamic spiral test	0.16
Number of changes in acceleration of dynamic spiral test	0.11
Mean radial velocity (static test)	0.27
Standard deviation radial velocity	0.35
Mean radial velocity (dynamic test)	0.25
Standard deviation radial velocity (dynamic test)	0.30
In-air time for stability test	0.22

It also demonstrates that newly introduced kinematic features related to radial velocity and in-air time for stability test have significant contribution to accurate diagnosis of PD in early stage and they are also flexible for the type of test. So, any combination of drawing test can be useful for diagnosis of disease (see Figures 3.4 and 3.5).

3.2.2.2.3 Histogram of Gradient Features

Spiral drawing dataset II consists of spiral and wave drawings of PD patients and healthy controls drawn using digitized tablet; spiral samples are shown in Figures 3.6 and 3.7, and wave samples are shown in Figures 3.8 and 3.9. The features are directly extracted from these drawings using a feature descriptor used to detect objects in computer vision and image processing. This technique

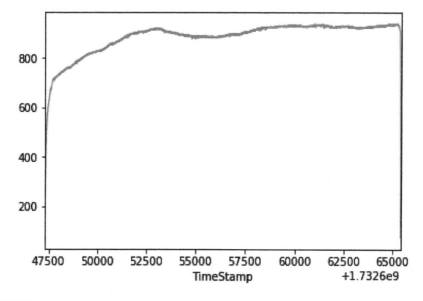

FIGURE 3.4 Spiral drawing stroke—healthy control.

Parkinson's Disease Imaging Diagnosis Using Digitized Spiral Drawing

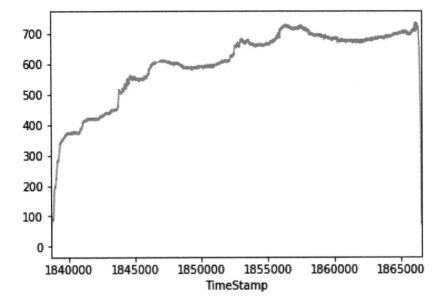

FIGURE 3.5 Spiral drawing stroke—PD patient.

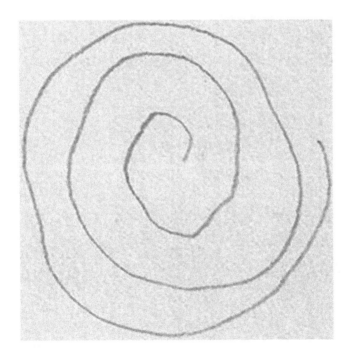

FIGURE 3.6 Patient spiral.

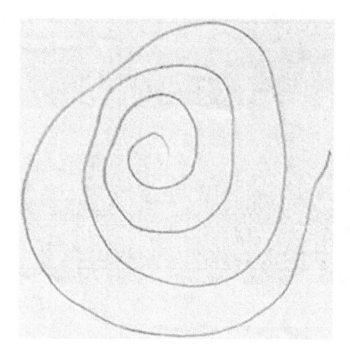

FIGURE 3.7 Healthy control spiral.

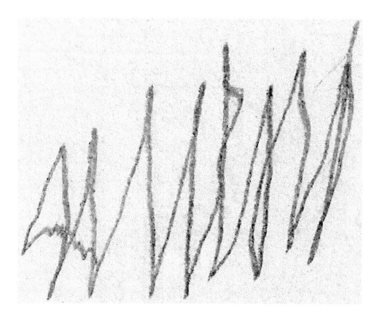

FIGURE 3.8 Patient wave pattern.

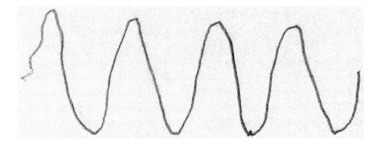

FIGURE 3.9 Healthy control wave pattern.

counts occurrences of gradient orientation in localised portions of an image. The generic steps are as follows:

- Partition the image into small connected regions called cells, and for each cell figure out a histogram of gradient directions. It can also be edge orientations by the processing pixels within the small cell.
- According to the gradient orientation, discrete points are converted to angular bins.
- These pixels contribute weighted gradient to the matching angular bin.
- Groups of nearby cells are treated as spatial regions named blocks. To normalise the histogram, the grouping of cells is the base. These histograms are treated as features for ML models.

The proposed work has applied the above feature extraction steps using skimage library and L1 normalisation. After normalisation and feature extraction, images are labelled as H for healthy control and P for PD patients (see Figure 3.10).

```
# compute the histogram of oriented gradients feature vector for
# the input image
features = feature.hog(image, orientations=19,
pixels_per_cell=(100,100), cells_per_block=(5,5),transform_
sqrt=True, block_norm="L1")

# return the feature vector
```

There are three significant methods for image data feature extraction using python.

> Method 1: Grayscale pixel values as features. In this method raw pixel values are treated as separate features. Feature vector is adding every value one after another. The quantifying values are intensity of brightness values of the pixels and will play the role of features. These values are in the form of a matrix of vectors. The number of features is the same as the number of pixels in the image. ML models can be directly applied on these raw features.
>
> Method 2: Mean pixel value of grayscale or colour scale. For grayscale images, the number of pixels is multiplied by 1 for one channel, and for colour images, the number of pixels is multiplied by 3 channels, as

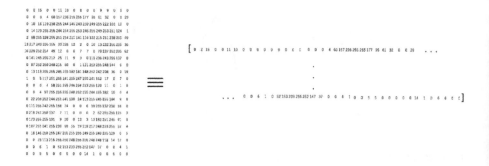

FIGURE 3.10 Intermediate grayscale feature extraction of wave pattern.

three colour values are the total features. Using matrix functions, these 3D matrix features are converted to single dimensional feature vector and ML models are applied on this feature dataset.

Method 3: Extracting edge features. Mathematically kernel function represents edge pixels. Using kernel, edge-oriented pixels are converted to significant and high correlation features. In this manner, difference of interior pixel and edge-aligned pixel is calculated and converted to a required feature. This third method is refined with histogram of features where it is the alignment which will differentiate or classify the images with the help of training algorithms in simple or ensemble models.

3.2.2.3 Model Selection and Parameter Tuning

The algorithms selected for solving the classification problem are logistic regression and ensemble model random forest because of their high interpretability and their effectiveness to handle imbalanced datasets. The class weight in these models was adjusted to be inversely relative to class frequencies in the input data. Some up-to-date supervised machine learning algorithms linear SVC C-support vector classification has also been employed.

    A. *Logistic Regression.* This is a machine learning classification technique applied to forecast the possibility of outcome of categorical dependent variables. The dataset consists of a target variable as binary variable and so contains only two values: 1 or 0. The logistic regression model predicts the probability of (target=1) as a function of an independent variable. Prerequisite of logistic regression is a binary-dependent variable; independent variables are independent of each other or have much less correlation. The dataset consists of imbalanced classes, so the ratio is 25:15. When too many categories are there, grouping of it is required. With the help of visualisation, we can determine a good predictor column from the dataset.

The performance of logistic regression can be tuned with its parameter solver. A solver is kind of a mathematical function based on hypothesis. Say H(x) is a function which takes an input x and produces the likely estimated output value. This nature of the hypothesis function varies from a one-variable linear equation or a very complicated and long multivariate equation as per the problem. Function supports to find the best parameters among the larger dataset that give us the least error (also named as cost or loss function) in predicting the output. Linear classification supports the logistic regression through solver terms. Each solver tries to find the parameter weights that minimise the cost. A coordinate descent (CD) algorithm for optimisation problems is the basis of solver. It successively performs approximate minimisation along coordinate axes or hyperplanes. The default solver is Liblinear, which applies automatic parameter selection. This is much slower, and with drawbacks of getting stuck in data processing. The proposed work is implementing lbfgs solver, which is much faster and produces the same accuracy. Solver lbfgs handles multinomial loss, with L2 regularisation with no penalty. In L2 regularisation, the sum of square of all feature weights is calculated by keeping the weights small but not zero, and so it is good for dense solution just like ridge regression. Lbfgs solver stores only the last few updates, so it also saves memory. Other advantages of lbfgs are that it is relatively fast and doesn't require similarly scaled data. It's the best choice for most cases without a really large dataset. So, the proposed work implements this logistic regression with lbfgs solver.

B. *Random Forest Classifier.* This is an ensemble algorithm. *Ensembled algorithms* combine more than one algorithm of same or different kind for classifying objects and then take votes for final consideration of class for test object.

The base classifier of random forest is decision tree. Random forest classifier creates a set of decision trees with random selection of subset of training set. It then aggregates the votes from different decision trees to decide the final class of the test object. The proposed work is implementing Random forest classifier with all default values of scikit-learn ensemble model, with 100 number of trees in the forest, gini impurity measure, minimum number of samples required to split the node is two, and all the features of dataset are considered for classification.

C. *Linear SVC C-Support Vector Classification.* Support vector classifier works on the principle of finding the best hyperplane to separate the different classes. It tries to maximise the distance between sample points and the hyperplane. The objective of a support vector classifier is to fit to the data to a 'best fit' hyperplane that classifies the data. C-support is the penalty parameter of the error term. It manages the trade-off between even decision boundary and the given training points correctly. It works on Rbf kernel function, which is a representation of nonlinear hyperplane. It is normally used in support vector machine classification and formulated as the squared Euclidean distance between the two feature vectors. The results of SVC also depend on gamma parameter for nonlinear hyperplanes. The higher the gamma value, the more likely the training dataset will fit the classifier. The proposed work is implementing default form of SVC from the scikit-learn library.

3.2.2.4 Experimental Setup

Python scikit-learn library models implemented are as follows:

- Logistic regression
- Random forest classifier
- Linear SVC

These models are then compared for the performance measures accuracy, precision, recall and F1 score.

Logistic regression and random forest classifier are applied on Spiral Drawing Dataset I, and Linear SVC and random forest classifier are implemented on Spiral Drawing Dataset II. First, feature engineering is done for kinematic features, and second, dataset feature engineering is done on images using HOG feature descriptor.

3.3 RESULTS

3.3.1 Metrics

tp:true positive, tn:true negative fp:false positive, fn=false negative are the notations.

- Accuracy: Classification accuracy is the same as the generic term accuracy. It is the ratio of the number of accurate predictions over the total number of given input samples.

$$Accuracy = Correctly classified instances\ /\ Total instances \qquad (3.2)$$

- Precision: Also called positive predictive value, this is the fraction of relevant instances among the retrieved instances. The precision is intuitively the ability of the classifier not to label a sample as positive if it is negative. The higher the value, the better the classifier.

$$Precision = tp\ /\ (tp + tn + fp + fn) \qquad (3.3)$$

- Sensitivity (Recall) is the fraction of the total amount of relevant instances that were actually retrieved.
  The recall is intuitively the ability of the classifier to find all the positive samples. The higher the value, the higher the accuracy.

$$Recall = tp/(tp + fn) \qquad (3.4)$$

- Specificity (also called true negative rate): Specificity relates to the classifier's ability to identify negative results
- F1 score: The F1 score is the $2*((precision*recall)/(precision + recall))$.

Parkinson's Disease Imaging Diagnosis Using Digitized Spiral Drawing

It is also called the F score or the F measure. When it is F1, it conveys the balance between the precision and the recall.

$$F1 = (2*(tp/(tp+fp+tn+fn))*(tp/(tp+fn)))/((tp/(tp+fp+tn+fn))+(tp/(tp+fn)))$$
$$(3.5)$$

It is observed that the Ensemble model gives better results as compared to logistic regression in the case of Dataset I with CSV files. The results are listed in Table 3.2 and Table 3.3. Linear SVC and ensemble models are producing comparable results with Dataset II drawing files. The results are listed in Table 3.4 and Table 3.5.

Time of execution is CPU time spent in the execution of the model, and mean iteration time is statistically relevant and calculated by executing the model at least for 10,000 iterations. Depending on feature extraction mechanism and parameter tuning of model, this may vary for every turn.

TABLE 3.2
Performance Metric I Evaluation. (Dataset I CSV files)

Model Applied	Precision	Recall	F1Score	Accuracy%
Logistic Regression	0.66	1.0	0.8	91.6
Random Forest Classifier	0.75	0.9	0.81	91.6

TABLE 3.3
Performance Metric II Evaluation. (Dataset I CSV files)

Model Applied	Time of Execution	Mean Iteration Time	Average Precision	AUC PR Curve
Logistic Regression	36 ms	851 µs	0.97	0.98
Random Forest Classifier	852 ms	1.12 ms	0.9	0.95

TABLE 3.4
Performance Metric I Evaluation. (Dataset II PNG files)

Model Applied	Precision	Recall	F1Score	Accuracy%
Linear SVC	0.36	0.47	0.58	73.3
Random Forest Classifier	0.37	0.9	0.48	83.3

TABLE 3.5

Accuracy Metric Evaluation. (Dataset II PNG files 30 number of images)

	Correctly Identified Images	Wrongly Identified Images
Random forest classifier	25	5
Linear SVC	22	8

The performance metric of precision and recall and F1score may give misleading results in the case of imbalanced dataset, as both are given equal importance in F1 score. So, for an accurate validation of the model, average precision value and precision recall Curve (PR curve) are also evaluated.

3.4 DISCUSSION

Precision metric demonstrates the rightness of model when it gives correct results where F1 score is the harmonic mean of precision and recall. The higher the values of precision, recall and F1 score, the better the validation of the model.

Also. the higher numerical value AUC for PR curve represents correctness of the model for classification.

Following is the code segment for parameter tuning for the HOG feature extractor and Ensemble model:

```
RandomForestClassifier(bootstrap=True, class_weight=None,
criterion='gini', oob_score=False, random_state=None,
verbose=0,
 warm_start=False
 max_depth=None, max_features='auto', max_leaf_nodes=None,
min_impurity_decrease=0.0, min_impurity_split=None, min_
samples_leaf=1, min_samples_split=2, min_weight_fraction_
leaf=0.0, n_estimators=100, n_jobs=None,
 )

model = RandomForestClassifier(n_estimators=100)model.
fit(trainX, trainY)
```

The proposed work is implemented on imbalanced dataset and training and test datasets are generated randomly. The results demonstrate better performance of logistic regression with relatively less time complexity as well. RF classifier also demonstrates better results but the computation time is high because of number of estimators, large number of features and optimum parameter values.

Following is the code segment for parameter tuning for HOG feature extractor and linear SVC:

```
from sklearn.svm import LinearSVC
from sklearn.datasets import make_classification
```

Parkinson's Disease Imaging Diagnosis Using Digitized Spiral Drawing   55

```
m1 = LinearSVC(random_state=0, tol=1e-7)
m1.fit(trainX, trainY)
LinearSVC(C=1.0, class_weight=None, dual=True,
verbose=0,loss='squared_hinge',fit_intercept=True,
intercept_scaling=1, max_iter=1000,
multi_class='ovr', penalty='l2', random_state=0, tol=1e-07,)
```

3.5 CONCLUSION

This chapter confirms that three types of digitized spiral drawing tests have a major impact on the classification of PD patients and healthy controls, when four ML models are implemented on mathematically processed datasets. The results are taken out on small-sized balanced and imbalanced datasets. Imbalanced dataset Spiral Drawing Dataset I consists of 40 persons' spiral drawing and CSV files. Balanced dataset Spiral Drawing Dataset II consists of 30 persons' spiral and wave pattern drawings. Dataset I feature engineering is done with the mathematical model and Dataset II feature engineering is done with HOG. Machine learning models are applied to classify the persons as patients and controls, and results are demonstrated with performance metrics.

Accuracy of both the mechanisms are comparable. The future work is determining severity of disease by applying fine-tuned ML models. Hence with the support of extended datasets and extended computational models, future diagnosis of PD can be done to support healthcare research for neurodegenerative diseases.

REFERENCES

1. Impedovo, Donato, et al. (2018). 'Dynamic handwriting analysis for supporting earlier Parkinson's disease diagnosis'. Information. 9(10): 247.https://doi.org/10.3390/info9100247.
2. Drotar, Peter, et al. (August 2014). 'Analysis of in-air movement in handwriting: A novel marker for Parkinson's Disease'. Comput. Meth. Prog. Bio. 117(3): 405–411.
3. Drotar, Peter, et al. (2016). 'Evaluation of handwriting kinematics and pressure for differential diagnosis of Parkinson's disease'. Artif. Intell. Med. 67: 39–46.
4. Impedovo, D., and Pirlo, G. (2019). 'Dynamic handwriting analysis for the assessment of neurodegenerative diseases: a pattern recognition perspective'. IEEE Rev. Biomed. Eng.12: 209–220.doi: 10.1109/RBME.2018.2840679.
5. Zham, Poonam, et al. (2019). 'A kinematic study of progressive micrographia in Parkinson's disease'. Fronti. Neurol. 10(403): 1–8. DOI=10.3389/fneur.2019.00403.
6. Eskidere, Omer, et al. (2012). 'A comparison of regression methods for remote tracking of Parkinson's disease progression'. Expert Syst. Appl. 39: 5523–5528.
7. Isenkul, Muhammed Erdem, et al. (2014). 'Improved spiral test using digitized graphics tablet for monitoring Parkinson's disease'. Proc. of ICEHTM.171–175.
8. San Luciano, M., et al. (2016). 'Digitized spiral drawing: A possible biomarker for early Parkinson's disease'. PLoS ONE. 11(10): e0162799.doi:10.1371/journal.pone. 0162799.
9. Aghanavesi, S., et al. (2017). 'Verification of a method for measuring Parkinson's disease related temporal irregularity in spiral drawings'. Sensors. 17(10): pii: E2341. doi: 10.3390/s17102341.
10. Miralles, F., et al. (2006). 'Quantification of the drawing of an Archimedes spiral through the analysis of its digitized picture'. J. Neurosci. Methods. 152:18–31.

11. Saunders-Pullman, R., Derby, C., Stanley, K., Floyd, A., Bressman, S., Lipton, R. B., et al. (2008). 'Validity of spiral analysis in early Parkinson's disease'. Mov. Disord. 23(4): 531–7. doi:10.1002/mds.21874.
12. Zham, P., Kumar, D. K., Dabnichki, P., Poosapadi Arjunan, S., and Raghav, S. (2017). 'Distinguishing different stages of Parkinson's disease using composite index of speed and pen-pressure of sketching a spiral'. Front. Neurol. 8:435. doi: 10.3389/fneur.2017.00435.
13. Thomas, Mathew, et al. (2017). 'Handwriting analysis in Parkinson's disease: Current status and future directions', Mov. Disord. (Wiley). 4(6):806–818.
14. Drotar, Peter, et al. (2015). 'Decision support framework for Parkinson's disease based on novel handwriting marker'. IEEE Trans. Neural Syst. Rehabil. Eng. 23(3):508–516.
15. João, W. M., et al. (2018). A new approach to diagnose Parkinson's disease using a structural co-occurrence matrix for a similarity analysis, Article ID 7613282, 8 pages, https://doi.org/10.1155/2018/7613282 Computational Intelligence and Neuroscience, 1687-5265, Hindawi.

4 Modelling and Analysis for Cancer Model with Caputo to Atangana-Baleanu Derivative

Ashish Mishra[1], Jyoti Mishra[2] and Vijay Gupta[3]
[1]Department of Computer Science and Engineering
Gyan Ganga Institute of Technology and Sciences
Jabalpur, Madhya Pradesh, India
[2]Department of Mathematics
Gyan Ganga Institute of Technology and Sciences
Jabalpur, Madhya Pradesh, India
[3]Department of Mathematics
University Institute of Technology-RGPV
Bhopal, Madhya Pradesh, India

CONTENTS

4.1 Introduction 57
4.2 Some Useful Definitions of Fractional Differential
Operators 58
4.3 Cancer Model 59
    4.3.1 Existence of Solution 59
    4.3.2 AB Derivative for Mathematical Cancer Model 62
    4.3.3 Numerical Solution for Mathematical Model Having Kernel
    Mittag-Leffler 66
4.4 Conclusions 69
References 69

4.1 INTRODUCTION

The future behaviour of tumour growth with the help of soft computing models and mathematical models have been broadly studied. The objective of this study is how tumours grow, how to discover the symptoms, how to detect unhealthy tissues and how we can predict the future condition of the disease. Chemotherapy, radiation and surgery are some treatment methods to increase cancer survival rates and care for the immune system [1–6]. In recent years, many people are doing research in cancer treatment and prediction. Mathematical models and soft computing play an important role in this field.

Mathematical models are commonly used among researchers for cancer treatment and further study, and researchers have been working in this field for the last five decades. Radiation helps to damage bad cells. Fractional calculus with fractional mathematical models helps the researcher to solve real-world problems with proper understanding. In addition, this numerical method provides essential information for the system [7, 8]. Some mathematical models square measure developed antecedently for cancer treatment. One of the treatment methods radiation provides to the body is the substitution of cells. In this chapter, we consider the three most popular fractional derivatives: C, CF and AB. Altogether we consider one tumour treatment mathematical model.

4.2 SOME USEFUL DEFINITIONS OF FRACTIONAL DIFFERENTIAL OPERATORS

Definition (i):

$$
{}_{0}^{C}D_t^\alpha[f(t)] = \frac{M(\alpha)}{(1-\alpha)} \int_0^t \frac{d}{dy} f(y) \exp\left[(t-y)\left(-\frac{\alpha}{1-\alpha}\right)\right] dy, \tag{4.1}
$$

Definition (ii):

$$
\frac{1}{\Gamma(n-\upsilon)} \int_a^t \frac{f^{(n)}(r)}{(t-r)^{\upsilon+1-n}} = {}_{a}^{C}D_t^\upsilon[f(t)], \ n > \upsilon > n-1 \in N \tag{4.2}
$$

Definition (iii):

$$
\int_a^t f(r)(t-r)^{\upsilon-1} dr \times \frac{1}{\Gamma(\upsilon)} = I^\upsilon[f(t)]. \tag{4.3}
$$

Definition (iv):

$$
{}^{CF}_{a}D_t^\upsilon[f(t)] = \frac{\upsilon M(\upsilon)}{(1-\upsilon)} \int_a^t \frac{d}{dx} \exp\left[(t-x)\times\left(-\frac{\upsilon}{1-\upsilon}\right)f(x)\right] dx, 0 < \alpha < 1 \tag{4.4}
$$

Definition (v):

$$
{}^{CF}I_t^\upsilon[f(t)] = \frac{2(1-\upsilon)}{2M(\upsilon)-\upsilon M(\upsilon)} u(t) + \frac{2\upsilon}{2M(\upsilon)-\upsilon M(\upsilon)} \int_0^t u(s)ds, \quad t \geq 0 \tag{4.5}
$$

Definition (vi): The ABC derivative is given by [6],

$$
{}^{ABC}_{0}D_t^\gamma[f(t)] = \frac{B(\gamma)}{(n-\gamma)} \int_0^t f^n(\theta) \, E_\gamma[-(t-\theta)^\gamma]d\theta, \tag{4.6}
$$

Analysis for Cancer Model with Caputo to Atangana-Baleanu Derivative 59

Definition (vii): by using the Laplace transform from both sides of equation (4.7), we get the following:

$$L\left\{{}^{ABC}_{0}D^{\gamma}_{t}[f(t)]\right\} = \frac{B(\gamma)}{(n-\gamma)}L\left[\int_0^t f^n(\theta)\,E_{\gamma}\left[-\frac{\gamma}{n-\gamma}(t-\theta)^{\gamma}\right]d\theta\right](s),$$

$$= \frac{B(\gamma)}{(n-\gamma)}\frac{s^{\gamma}L[f(t)](s)-s^{\gamma-1}f(0)}{s^{\gamma}+\dfrac{\gamma}{1-\gamma}}L\left[\int_0^t f^n(\theta)\,E_{\gamma}\left[-\frac{\gamma}{n-\gamma}(t-\theta)^{\gamma}\right]d\theta\right](s),$$

(4.7)

Definition (viii): The definition of the AB derivative is given by:

$${}^{ABC}_{0}D^{\gamma}_{t}[f(t)] = \frac{B(\gamma)}{(n-\gamma)}\int_0^t f^n(\theta)\,E_{\gamma}\left[-\frac{\gamma}{n-\gamma}(t-\theta)^{\gamma}\right]d\theta,\quad n-1<\gamma\le n,\quad (4.8)$$

4.3 CANCER MODEL

The cancer is given as:

$$-\varepsilon D(t)\eta + \alpha_1\eta\left(1-\frac{\eta}{S_1}\right) - \beta_1\eta a = \frac{d\eta(t)}{dt}$$

$$-a\times D(t)+\alpha_2 a\left(1-\frac{\eta}{S_2}\right) - \beta_2\eta a = \frac{da(t)}{dt}$$

(4.9)

The strategy of radiation is denoted by $D(t)$. It is assumed that $\gamma > 0$, treatment stage when $t \in [n\omega, n\omega + L]$, no treatment stage when $t \in [n\omega + L, (1+n)\omega]$ $\forall$ $n = 0,1,2,\dots$.

where
ω Radiotherapy. The given system (4.9) is given by:

$$^{CF}_{0}D^{\upsilon}_{t}[\eta] = \eta\left(1-\frac{\rho}{S_1}\right)\alpha_1 - \beta_1\eta a - \varepsilon D(t)\eta$$

$$^{CF}_{0}D^{\upsilon}_{t}[a] = a\left(1-\frac{\eta}{S_2}\right)\alpha_2 - \beta_2\eta a - D(t)a$$

(4.10)

4.3.1 EXISTENCE OF SOLUTION

For this, consider the integral equations in the following manner:

$$\frac{2(1-\upsilon)}{2M(\upsilon)-\upsilon M(\upsilon)}\left[\alpha_1\eta\left(1-\frac{\eta}{S_1}\right)-\beta_1\eta a-\varepsilon D(t)\eta\right]+$$

$$\frac{2\upsilon}{2M(\upsilon)-\upsilon M(\upsilon)}\int_0^t\left[\alpha_1\eta\left(1-\frac{\eta}{S_1}\right)-\beta_1\eta a-\eta\varepsilon D(t)\right]=\eta(t)-\eta_0(t)$$

$$\frac{2(1-\upsilon)}{2M(\upsilon)-\upsilon M(\upsilon)}\left[\alpha_2 a\left(1-\frac{\eta}{S_2}\right)-\beta_2\eta a-\varepsilon D(t)a\right]+$$

$$\frac{2\upsilon}{2M(\upsilon)-\upsilon M(\upsilon)}\int_0^t\left[-\varepsilon D(t)a+\alpha_2 a\left(1-\frac{\eta}{S_2}\right)-\beta_2\eta a\right]=a(t)-a_0(t)$$

(4.11)

Considering the fractional integral of order υ, we get

$$\frac{2(1-\upsilon)}{2M(\upsilon)-\upsilon M(\upsilon)}\left[\alpha_1\eta\left(1-\frac{\eta}{S_1}\right)-\beta_1\eta a-\varepsilon D(t)\eta\right]+$$

$$\frac{2\upsilon}{2M(\upsilon)-\upsilon M(\upsilon)}\int_0^t\left[\alpha_1\eta\left(1-\frac{\eta}{S_1}\right)-\beta_1\eta a-\eta\varepsilon D(t)\right]=\eta-\eta_0$$

Let us suppose that the kernels are defined by:

$$\alpha_1\eta\left(1-\frac{\eta}{S_1}\right)-\eta a\beta_1-\varepsilon D(t)\eta=s(\eta(t),t)$$

$$\alpha_2 a\left(1-\frac{\eta}{S_2}\right)-\eta a\beta_2-D(t)a=s(a(t),t)$$

Let us consider the operator T is a mapping from H to H that is compact. Then we get

$$T\eta(t)=\frac{2(1-\upsilon)}{2M(\upsilon)-\upsilon M(\upsilon)}s(t,\eta(t)+\frac{2\upsilon}{2M(\upsilon)-\upsilon M(\upsilon)}\int_0^t s(t,\eta(t)dy$$

(4.12)

$$Ta(t)=\frac{2(1-\upsilon)}{2M(\upsilon)-\upsilon M(\upsilon)}s(t,\eta(t)+\frac{2\upsilon}{2M(\upsilon)-\upsilon M(\upsilon)}\int_0^t s(t,\eta(t)dy$$

Theorem 1: If T is compact, then it is completely continuous
Proof: Suppose the $H \supset M$ is bounded. There exist $\|\rho\| < l$ and $\|a\| < m$

$$|T\eta(t)|\le\frac{2L_1}{2M(\upsilon)-\upsilon M(\upsilon)}[1-\upsilon+\upsilon c_1]$$

Analysis for Cancer Model with Caputo to Atangana-Baleanu Derivative

Similarly, we have

$$|Ta(t)| \le \left[\frac{2(1-\upsilon)}{2M(\upsilon)-\upsilon M(\upsilon)} + \frac{2\upsilon}{2M(\upsilon)-\upsilon M(\upsilon)} c_2 \right] |L_2$$

$$|Ta(t)| \le \frac{2L_2}{2M(\upsilon)-\upsilon M(\upsilon)} [1-\upsilon+\upsilon c_2]$$

Therefore, we conclude that T(M) is bounded.

Now we consider $t_1 < t_2$ and $\eta(t)$, $a(t) \in M$ and there exist $\epsilon > 0$ provided that $|t_1 - t_2| < \delta$.

We have

$$\|T\eta(t_2)-T\eta(t_1)\| \le \left| \frac{2(1-\upsilon)}{2M(\upsilon)-\upsilon M(\upsilon)} ((s(t_2,\eta(t_2)) - (s(t_1,\eta(t_1)))) \right|$$

$$+ \left| \frac{2\upsilon}{2M(\upsilon)-\upsilon M(\upsilon)} \int_0^{t_2} s(t,\eta(t)dy - \frac{2\upsilon}{2M(\upsilon)-\upsilon M(\upsilon)} \int_0^{t_1} s(t,\eta(t)dy \right|$$

$$\le \frac{2(1-\upsilon)}{2M(\upsilon)-\upsilon M(\upsilon)} |((s(t_2,\eta(t_2)) - (s(t_1,\eta(t_1))))| \tag{4.13}$$

$$+ L_1 |((s(t_2,\eta(t_2)) - (s(t_1,\eta(t_1))))| \times \frac{2\upsilon}{2M(\upsilon)-\upsilon M(\upsilon)}$$

Now we investigate the following part:

$$\le c_3\lambda - c_4\lambda - c_5\lambda \text{ where } \lambda = |(\eta(t_2)-\eta(t_1))| $$
$$\le C|t_2 - t_1| \tag{4.14}$$

Now from equation (4.16) and (4.17), we have

$$|T\eta(t_2)-T\eta(t_1)| \le \frac{2(1-\upsilon)}{2M(\upsilon)-\upsilon M(\upsilon)} C|t_2 - t_1| + \frac{2(1-\upsilon)}{2M(\upsilon)-\upsilon M(\upsilon)} L_1 |t_2 - t_1|$$

$$\delta = \frac{\epsilon}{\dfrac{2(1-\upsilon)}{2M(\upsilon)-\upsilon M(\upsilon)} C + \dfrac{2(1-\upsilon)}{2M(\upsilon)-\upsilon M(\upsilon)} L_1} \tag{4.15}$$

$$\text{Provided that } |T\iota(t_2)-T\iota(t_1)| \le \epsilon \tag{4.16}$$

Now similarly with help of the condition we get

$$\delta = \frac{\epsilon}{\dfrac{2(1-\upsilon)}{-\upsilon M(\upsilon)+2M(\upsilon)} D + \dfrac{2(1-\upsilon)}{-\upsilon M(\upsilon)+2M(\upsilon)} L_2}$$

62 Soft Computing Applications and Techniques in Healthcare

Such that $|Tc(t_2) - Tc(t_1)| \leq \varepsilon$. Hence T(M) is continuous.

Uniqueness Solution:

$$\left| \frac{2\upsilon}{2M(\upsilon) - \upsilon M(\upsilon)} (s(t,\iota_1(t)) - s(t,\iota_2(t))) \right| + \left| \frac{2\upsilon}{2M(\upsilon) - \upsilon M(\upsilon)} \right| \left| \int_0^t (s(t,\iota_1(t)) - s(t,\iota_2(t)))dy \right|$$

$$= |T \times \iota_1(t) - T \times \iota_2(t)|$$

$$\leq \frac{2(1-\eta)}{2M(\eta) - \upsilon M(\eta)} F_1 |\iota_1(t) - \iota_2(t)| + \frac{2\eta}{2M(\eta) - \eta M(\eta)} F_1 |\iota_1(t) - \iota_2(t)|$$

Hence the mapping T is a contradiction if the following conditions satisfied:

$$\left\{ \frac{2(1-\eta)}{2M(\eta) - \eta M(\eta)} F_2 + \frac{2\eta}{2M(\eta) - \eta M(\eta)} F_2 \right\} < 1 \tag{4.17}$$

Therefore, we conclude that the model has a single solution.

4.3.2 AB Derivative for Mathematical Cancer Model

Taking into consideration the equation for Mittag-Leffler Kernel can be given by:

$$-\varepsilon D(t)\rho + \alpha_1 \rho \left(1 - \frac{\rho}{S_1} \right) - \beta_1 \rho a = {}^{ABC}_0 D_t^\gamma [\rho(t)]$$

$$-D(t)a + \alpha_2 a \left(1 - \frac{a}{S_2} \right) - \beta_2 \rho a = {}^{ABC}_0 D_t^\gamma [a(t)] \tag{4.18}$$

First we use the fixed-point theorem. Now writing the integral transform equation of (4.19), we get

$$\frac{(1-\upsilon)}{B(\upsilon)} \left[\left(1 - \frac{\rho}{S_1} \right) \alpha_1 \rho - \beta_1 \rho a - \varepsilon D(t)\rho \right] +$$

$$\frac{\upsilon}{B(\upsilon)\,\Gamma(\upsilon)} \int_0^t \left[\alpha_1 \rho \left(1 - \frac{\rho}{S_1} \right) - \beta_1 \rho a - \varepsilon D(y)\rho \right] dy = \rho(t) - \rho_0(t). \tag{4.19}$$

With order υ, we get

Analysis for Cancer Model with Caputo to Atangana-Baleanu Derivative 63

$$\frac{(1-\upsilon)}{B(\upsilon)}\left[\alpha_1\rho\left(1-\frac{\rho}{S_1}\right)-\beta_1\rho a-\varepsilon B(t)\rho\right]+$$

$$\frac{\upsilon}{B(\upsilon)\,\Gamma(\upsilon)}\int_0^t\left[\alpha_1\rho\left(1-\frac{\rho}{S_1}\right)-\beta_1\rho a-\varepsilon B(y)\rho\right]dy=\rho(t)-\rho_0(t)$$

$$\frac{(1-\upsilon)}{B(\upsilon)}\left[\alpha_2 a\left(1-\frac{a}{S_2}\right)-\beta_2\rho a-\varepsilon B(t)a\right]+$$

$$\frac{2\upsilon}{B(\upsilon)\,\Gamma(\upsilon)}\int_0^t\left[\alpha_2 a\left(1-\frac{a}{S_2}\right)-\beta_2\rho a-\varepsilon B(y)a\right]dy=c(t)-c_0(t)$$

$$s(t,\delta(t)=\alpha_1\delta\left(1-\frac{\delta}{S_1}\right)-\beta_1\delta a-\varepsilon D(t)\delta$$

$$s(t,c(t)=\alpha_2 a\left(1-\frac{\delta}{S_2}\right)-\beta_2\delta a-D(t)a$$

Let us consider the mapping T from H to H. Now we show that this operator is compact. In that case we get

$$T\rho(t)=\frac{(1-\upsilon)}{B(\upsilon)}s(t,\rho(t)+\frac{\upsilon}{B(\upsilon)\Gamma(\upsilon)}\int_0^t s(t,\rho(t)dy$$

$$(4.20)$$

$$Ta(t)=\frac{(1-\upsilon)}{B(\upsilon)}s(t,\rho(t)+\frac{\upsilon}{B(\upsilon)\Gamma(\upsilon)}\int_0^t s(t,\rho(t)dy$$

Theorem 2: If $T:H\to H$ is compact, then it is completely continuous

Proof: Suppose that M is a subset of H which is bounded $\exists\|\rho\|<l$ and $\|a\|<m$ where $l,\,m<0$:

$$\max_{\substack{0<t<1\\0\le\rho\le l}} s(t,\rho(t)=L_3 \text{ and } L_4=\max_{\substack{0<t<1\\0\le a\le m}} s(t,a(t) \; \forall\rho,a\in M$$

$$|T\rho(t)|=\left|\frac{(1-\upsilon)}{B(\upsilon)}s(t,\rho(t)+\frac{2\upsilon}{B(\upsilon)\Gamma(\upsilon)}\int_0^t s(t,\rho(t)dy\right|$$

$$|T\rho(t)|\le\left|\frac{(1-\upsilon)}{B(\upsilon)}\right|s(t,\rho(t)+\left|\frac{\upsilon}{B(\upsilon)\Gamma(\upsilon)}\right|\left|\int_0^t s(t,\rho(t)dy\right|$$

$$|T\rho(t)|\le\left[\frac{(1-\upsilon)}{B(\upsilon)}+\frac{\upsilon}{B(\upsilon)\Gamma(\upsilon)}c_1\right]|L_3$$

$$|T\rho(t)|\le\frac{L_3}{B(\upsilon)}\left[1-\upsilon+\frac{\upsilon c_1}{\Gamma(\upsilon)}\right]$$

in the same way,

$$|Ta(t)| \leq \left|\frac{(1-\upsilon)}{B(\upsilon)}\right| s(t,a(t) + \left|\frac{\upsilon}{\Gamma\upsilon M(\upsilon)}\right| \left|\int_0^t s(t,a(t)dy\right|$$

$$|Ta(t)| \leq \frac{L_4}{B(\upsilon)}\left[(1-\upsilon) + \frac{c_2\upsilon}{\Gamma\upsilon}\right]$$

Thus, we conclude that T(M) is bounded.

At the present we suppose $t_1 < t_2$ and $\rho(t)$, $a(t) \in M$ and there exists $\in > 0$ provided that $|t_1 - t_2| < \delta$.

We have

$$\left\|T\rho(t_2) - T\rho(t_1)\right\| \leq \left|\frac{(1-\upsilon)}{B(\upsilon)}((s(t_2,\rho(t_2)) - (s(t_1,\rho(t_1)))\right|$$

$$+\left|\frac{\upsilon}{B(\upsilon)\Gamma(\upsilon)}\int_0^{t_2} s(t,\rho(t)dy - \frac{\upsilon}{B(\upsilon)\Gamma(\upsilon)}\int_0^{t_1} s(t,\rho(t)dy\right| \quad (4.21)$$

$$\left\|T\rho(t_2) - T\rho(t_1)\right\| \leq \frac{(1-\upsilon)}{B(\upsilon)}\left|((s(t_2,\rho(t_2)) - (s(t_1,\rho(t_1)))\right|$$

$$+\frac{\upsilon}{B(\upsilon)\Gamma(\upsilon)}L_1\left|((s(t_2,\rho(t_2)) - (s(t_1,\rho(t_1)))\right|$$

Now we examine the following part:

$$\leq c_3\left|(\rho(t_2) - \rho(t_1))\right| - c_4\left|(\rho(t_2) - \rho(t_1))\right| - c_5\left|(\rho(t_2) - \rho(t_1))\right|$$

$$\leq (c_3 - c_4 - c_5)\left|(\rho(t_2) - \rho(t_1))\right| \quad (4.22)$$

$$\leq E|t_2 - t_1|$$

Now from equation (4.25) and (4.26), we have

$$|T\rho(t_2) - T\rho(t_1)| \leq \frac{(1-\upsilon)}{B(\upsilon)}E|t_2 - t_1| + \frac{(1-\upsilon)}{B(\upsilon)\Gamma(\upsilon)}L_3|t_2 - t_1|$$

$$\delta = \frac{\varepsilon}{\dfrac{(1-\upsilon)}{B(\upsilon)}E + \dfrac{(1-\upsilon)}{B(\upsilon)\Gamma\upsilon}L_3} \quad (4.23)$$

provided that

$$|T\rho(t_2) - T\rho(t_1)| \leq \varepsilon \quad (4.24)$$

Analysis for Cancer Model with Caputo to Atangana-Baleanu Derivative

Similarly we can get the following for the function a with the same conditions. Then we get

$$\delta = \frac{\varepsilon}{\dfrac{(1-\upsilon)}{B(\upsilon)}F + \dfrac{(1-\upsilon)}{B(\upsilon)\Gamma(\upsilon)}L_4} \tag{4.25}$$

such that $|Ta(t_2) - Ta(t_1)| \leq \varepsilon$. Hence T(M) is continuous.

Uniqueness Solution:

$$|T\rho_1(t) - T\rho_2(t)| = \left| \frac{\upsilon}{B(\upsilon)}(s(t,\rho_1(t)) - s(t,\rho_2(t))) + \frac{\upsilon}{B(\upsilon)\Gamma(\upsilon)} \right| \left| \int_0^t (s(t,\rho_1(t)) - s(t,\rho_2(t)))dy \right|$$

$$\leq \frac{(1-\upsilon)}{B(\upsilon)} |(s(t,\rho_1(t)) - s(t,\rho_2(t)))| + \left| \frac{\upsilon}{B(\upsilon)\Gamma(\upsilon)} \right| \left| \int_0^t (s(t,\rho_1(t)) - s(t,\rho_2(t)))dy \right|$$

$$\leq \frac{(1-\upsilon)}{B(\upsilon)} F_1 |\rho_1(t) - \rho_2(t)| + \frac{\upsilon}{B(\upsilon)\Gamma(\upsilon)} F_1 |\rho_1(t) - \rho_2(t)|$$

For simplicity we write

$$|T\rho_1(t) - T\rho_2(t)| \leq \left\{ \frac{(1-\upsilon)}{B(\upsilon)} F_1 + \frac{\upsilon}{B(\upsilon)\Gamma(\gamma)} F_1 \right\} |\rho_1(t) - \rho_2(t)|$$

Similarly for the function with a we get

$$|Tc_1(t) - Tc_2(t)| = \left| \frac{2\upsilon}{2M(\upsilon) - \upsilon M(\upsilon)}(s(t,c_1(t)) - s(t,c_2(t))) \right|$$

$$+ \left| \frac{2\upsilon}{2M(\upsilon) - \upsilon M(\upsilon)} \right| \left| \int_0^t (s(t,c_1(t)) - s(t,c_2(t)))dy \right|$$

$$\leq \frac{2(1-\upsilon)}{2M(\upsilon) - \upsilon M(\upsilon)} |(s(t,c_1(t)) - s(t,c_2(t)))| + \left| \frac{2\upsilon}{2M(\upsilon) - \upsilon M(\upsilon)} \right| \left| \int_0^t (s(t,c_1(t)) - s(t,c_2(t)))dy \right|$$

$$\leq \frac{2(1-\upsilon)}{2M(\upsilon) - \upsilon M(\upsilon)} F_2 |c_1(t) - c_2(t)| + \frac{2\upsilon}{2M(\upsilon) - \upsilon M(\upsilon)} F_2 |c_1(t) - c_2(t)|$$

$$\leq \left(\frac{2(1-\upsilon)}{2M(\upsilon) - \upsilon M(\upsilon)} F_2 + \frac{2\upsilon}{2M(\upsilon) - \upsilon M(\upsilon)} F_2 \right) |c_1(t) - c_2(t)|$$

66   Soft Computing Applications and Techniques in Healthcare

Hence the mapping T is a contradiction if the following conditions satisfied:

$$\left\{ \frac{2(1-\upsilon)}{2M(\upsilon)-\upsilon M(\upsilon)} F_2 + \frac{2\upsilon}{2M(\upsilon)-\upsilon M(\upsilon)} F_2 \right\} < 1 \tag{4.26}$$

Hence this is the complete solution of this numerical solution.

4.3.3 NUMERICAL SOLUTION FOR MATHEMATICAL MODEL HAVING KERNEL MITTAG-LEFFLER

Taking into consideration equation (4.12), the adapted epidemic representation is given by

$$\begin{aligned}
{}^{ABC}_{0}D^{\gamma}_{t}[\rho(t)] &= \alpha_1\rho\left(1-\frac{\rho}{S_1}\right) - \beta_1\rho a - \varepsilon D(t)\rho \\
{}^{ABC}_{0}D^{\gamma}_{t}[c(t)] &= \alpha_2 c\left(1-\frac{c}{S_2}\right) - \beta_2\rho c - D(t)c
\end{aligned} \tag{4.27}$$

${}^{ABC}_{0}D^{\gamma}_{t}$ AB fractional derivative and $0 < \gamma \le 1$

Applying the Sumudu transform operator in equation (4.32) to both sides, we get

$$\begin{aligned}
ST[\rho(t) - \rho(0)] &\times E_\mu\left(-\frac{1}{1-\mu}\right) \times \frac{B(\mu)\mu\Gamma(\mu+1)}{1-\mu} \\
&= ST\left[\alpha_1\rho\left(1-\frac{\rho}{S_1}\right) - \beta_1\rho a - \varepsilon D(t)\rho\right] \\
ST[c(t) - c(0)] &\times \frac{B(\mu)\mu\Gamma(\mu+1)}{1-\mu} \times E_\gamma\left(-\frac{1}{1-\mu}\right) \\
&= ST\left[\alpha_2 c\left(1-\frac{c}{S_2}\right) - \beta_2\rho c - D(t)c\right]
\end{aligned} \tag{4.28}$$

Rearranging equation (4.33), we obtain

$$\begin{aligned}
ST[\rho(t)] &= \rho(0) + \frac{1-\gamma}{E_\gamma\left(-\dfrac{1}{1-\gamma}u^\gamma\right) \times \gamma \times \Gamma(\gamma+1) \times B(\gamma)} \cdot \\
&\qquad ST\left[\alpha_1\rho\left(1-\frac{\rho}{S_1}\right) - \beta_1\rho a - \varepsilon D(t)\rho\right] \\
ST[a(t)] &= a(0) + \frac{1-\gamma}{E_\gamma\left(-\dfrac{1}{1-\gamma}u^\gamma\right) \gamma \times \Gamma(\gamma+1) \times B(\gamma)} \cdot \\
&\qquad ST\left[\alpha_1 a\left(1-\frac{\rho}{S_1}\right) - \beta_1\rho a - \varepsilon D(t)a\right]
\end{aligned} \tag{4.29}$$

Analysis for Cancer Model with Caputo to Atangana-Baleanu Derivative

$$\rho(t) = \rho(0) + ST^{-1}$$

$$\left\{ \frac{1-\gamma}{E_\gamma\left(-\frac{1}{1-\gamma}u^\gamma\right)\gamma \times \Gamma(\gamma+1) \times B(\gamma)} \cdot ST\left[\alpha_1\rho\left(1-\frac{\rho}{S_1}\right)-\beta_1\rho a - \varepsilon D(t)\rho\right]\right\}$$

(4.30)

$$a(t) = a(0) + ST^{-1}$$

$$\left\{ \frac{1-\gamma}{E_\gamma\left(-\frac{1}{1-\gamma}u^\gamma\right)\gamma \times \Gamma(\gamma+1) \times B(\gamma)} \cdot ST\left[\alpha_1 a\left(1-\frac{\rho}{S_1}\right)-\beta_1\rho a - \varepsilon D(t)a\right]\right\}$$

Now we obtain

$$\rho_{(n+1)}(t) = \rho_{(n)}(0) + ST^{-1}$$

$$\left\{ \frac{1-\gamma}{E_\gamma\left(-\frac{1}{1-\gamma}u^\gamma\right)\gamma \times \Gamma(\gamma+1) \times B(\gamma)} \cdot ST\left[\alpha_1\rho_{(n)}\left(1-\frac{\rho_{(n)}}{S_1}\right)-\beta_1\rho_{(n)}a - \varepsilon D(t)\rho_{(n)}\right]\right\}$$

(4.31)

$$a_{(n+1)}(t) = a_{(n)}(0) + ST^{-1}$$

$$\left\{ \frac{1-\gamma}{E_\gamma\left(-\frac{1}{1-\gamma}u^\gamma\right)\gamma \times \Gamma(\gamma+1) \times B(\gamma)} \cdot ST\left[\alpha_1 a_{(n)}\left(1-\frac{a_{(n)}}{S_1}\right)-\beta_1\rho_{(n)}a - \varepsilon D(t)a_{(n)}\right]\right\}$$

Then we have $\rho(t) = \lim_{n\to\infty}\rho_n(t)$; at the present, we give the stability investigation of this technique. Let Banach space be $(Y, |.|)$ as well as N is a mapping of Y. If $z_{n+1} = g(N, z_n)$, then the situation is written below has to be fulfilled for $Nz_n = z_{n+1}$

1. N should have at least one element of the fixed-point set z_n
2. $P = \lim_{n\to\infty}\rho_n(t)$.

Theorem 3: If H is a mapping on X and X is a Banach space, then
$$\|N_x - N_z\| \le \eta\|Y - N_x\| + \mu\|y - z\|$$

For all $y, z \in Y$, where $0 \le \eta < 1$.

Now with the help of equation (4.31) we have

$$\frac{1-\mu}{B(\mu)\mu\Gamma(\mu+1)E_\mu\left(-\dfrac{1}{1-\mu}u^\gamma\right)},$$

fractional Lagrange multiplier corresponds with the above equations.

Theorem 4: If K is a mapping, then

$$\rho_{(n)}(0) + ST^{-1}$$

$$\left\{\frac{1-\mu}{E_\mu\left(-\dfrac{1}{1-\mu}u^\mu\right)\gamma\times\Gamma(\gamma+1)\times B(\gamma)} \cdot ST\left[\alpha_1\rho_{(n)}\left(1-\frac{\rho_{(n)}}{S_1}\right)-\beta_1\rho_{(n)}a-\varepsilon D(t)\rho_{(n)}\right]\right\}$$

$$= K[\rho]$$

$$c_{(n)}(0) + ST^{-1}$$

$$\left\{\frac{1-\gamma}{E_\mu\left(-\dfrac{1}{1-\mu}u^\mu\right)\gamma\times\Gamma(\gamma+1)\times B(\gamma)} \cdot ST\left[\alpha_1 c_{(n)}\left(1-\frac{c_{(n)}}{S_1}\right)-\beta_1\rho_{(n)}c-\varepsilon D(t)c_{(n)}\right]\right\}$$

$$= K[c]$$

With the property of the norm and taking into consideration the triangular inequity, we get

$$\|\rho_n(t) - X(t)\| \ge \|K[\rho_n(t)] - K[X(t)]\|$$

$$+ST^{-1}\left\{\frac{1-\gamma}{B(\gamma)\gamma\Gamma(\gamma+1)E_\gamma\left(-\dfrac{1}{1-\gamma}u^\gamma\right)}\right\}.$$

$$
\eta = \begin{cases}
\left\| c_n(t) - Y_{2(m)}(t) \right\| \beta_2 \rho - \left\| c_n(t) - Y_{2(m)}(t) \right\| \varepsilon B(t) + \\[2mm]
\alpha_1 \left\| \rho_n(t) \right\| \times \alpha_2 \left\| c_n(t) \right\| \times \left\| \dfrac{-c_n(t) + Y_{2(m)}(t)}{S_2} \right\| + \\[4mm]
\left\| - X_{1(m)}(t) + \rho_n(t) \right\| \beta_1 c - \varepsilon B(t) + \alpha_1 \left\| \rho_n(t) - Y_{1(m)}(t) \right\| \times \left\| \dfrac{+Y_{1(m)}(t - \rho_n(t))}{S_1} \right\|
\end{cases}
$$

This is the complete solution.

4.4 CONCLUSIONS

In this chapter, we covered a cancer treatment model through the new fractional derivative which incorporates the contacts connecting well tissue cells, cancer cells and activation-resistant classification cells. The solutions of the decision models were obtain by numerical method – that is, an iterative numerical method. We have a tendency to find the uniqueness and existence solution of the tumour growth model. In this chapter we have shown how the solution provided the unique solution. This work will definitely help researchers who are working in tumour treatment and prediction of growth. The outcome offers the necessary and sufficient information regarding the proposed fractional derivatives.

REFERENCES

1. Mishra, J. (2018). 'A remark on fractional differential equation involving I-function'. Eur. Phys. J. Plus. 133(2): 36.
2. Losada, J., and Nieto, J. (2015). 'Properties of a new fractional derivative without singular kernel'. Progr. Fract. Differ. Appl. 1: 87.
3. Caputo, M. (1967). 'Linear models of dissipation whose Q is almost frequency independent-II geophys'. J. R. Astr. Soc. 13: 529–539.
4. Mishra, J. (2019). Numerical Analysis of a Chaotic Model with Fractional Differential Operators: From Caputo to Atangana–Baleanu Methods of Mathematical Modelling: Fractional Differential Equations, 167.
5. Mishra, J. (2019). 'Modified Chua chaotic attractor with differential operators with non-singular kernels'. Chaos Soliton. Fract. 125: 64–72.
6. Mishra, J. (2019). 'Analysis of the Fitzhugh-Nagumo model with a new numerical scheme'. Discrete Cont. Dyn-S. March 2020, 13(3): 781–795.
7. Mishra, J. (2018). 'Fractional hyper-chaotic model with no equilibrium'. Chaos Soliton. Fract. 116: 43–53.
8. Mishra, A. (2016). 'A unified problem related to a String and H – Function'. ISSN-0975-1726, Vol. 21, Page No.1–4.

5 Selection of Hospital Using Integrated Fuzzy AHP and Fuzzy TOPSIS Method

Vikas Shinde and Santosh K. Bharadwaj
Department of Applied Mathematics
Madhav Institute of Technology and Science
Gwalior, Madhya Pradesh, India

CONTENTS

5.1 Introduction 71
5.2 Preliminaries of Fuzzy Sets and Fuzzy Numbers 73
5.3 Analytic Hierarchy Process (AHP) 74
5.4 Fuzzy Analytic Hierarchy Process (FAHP) 75
    5.4.1 Chang's Extent Analysis of FAHP 76
5.5 Topsis Method.. 78
5.6 Selection Criteria of Hospital 80
5.7 Evaluation Framework 80
    5.7.1 Implication of Fuzzy AHP 81
    5.7.2 Implementation of TOPSIS 86
5.8 Conclusion 93
References 93

5.1 INTRODUCTION

Recently, several viruses have come into existence which impact multiple organs of the human body. This makes finding effective and suitable medical facilities more important than ever, whether the problem is life-threatening or involves major or minor surgery. This is an increasingly challenging issue for medical science. Some problems such as heart attack, and kidney, liver and other major disease require immediate care to avoid additional complications. Such care is only provided by major hospitals where a panel of specialists are available around the clock, along with trained staff. Nowadays it has been observed that major corporations open new well-equipped hospitals which are very expensive and often not affordable. Thus, people need to find good hospital facilities within their budgets. In this situation the patient and family members must take many factors into consideration to obtain better treatment in good hospitals. The fuzzy analytic hierarchy process (FAHP) is an important technique of multi-criteria decision making (MCDM), which has improved over five decades to analyse qualitative and quantitative study.

AHP is not capable of providing the innate ambiguity with the image of the judges' acuity to obtain accurate numbers. Thus, FAHP is used to adequately handle the fuzziness and vagueness because these are common characteristics of decision-making difficulties. It is an important part of MCDM techniques. MCDM is a powerful tool for modelling purpose as well as the methodology of dealing with critical problems. MCDM includes both multi-objective decision making (MODM) and multi-attribute decision making (MADM). MADM refers to making prior judgments among other existing alternatives which are depicted by multiple, conflicting attributes. However, MADM problems are routine problems of our life. For example, in buying a house, attributes such as cost, size, neighbourhood and closeness to market and hospital should be evaluated. In buying a car, factors such as cost, comfort, safety, fuel consumption and so on are considered. Simultaneously this technique is also used in organisations for planning issues, supplier selection, resource allocation, expert system design and so on. TOPSIS, VIKOR, ELECTRE, PROMETHEE, ORESTE and MAUT are other techniques of MADM apart from AHP.

Similarly, the selection of the best hospital is a MCDM problem which is affected by many conflicting components in order to make the appropriate decision. It has vast applicability of real-world engineering, scientific, medical and information technology problems wherein selection criteria required. Many researchers have paid attention to such studies. R. Andres et al. constructed fuzzy comparison matrices using linguistic variables for pair-wise comparison of each criterion [1]. In addition, H. Bronja used AHP in combination with TOPSIS to find the difficulty of criteria to get a ranking of alternatives [2]. D. Y. Chang invented a novel approach for fuzzy AHP with use of TFN for pair-wise comparison scale of fuzzy AHP [3]. S. Chou et al. developed subjective and objective attributes in addition to group decision for location and selection under multi-attribute decision making (MADM) problems [4]. Dagdevien and Yuksel introduced behaviour-based safety systems to assign the different weight-age in different attributes using fuzzy analytic hierarchy process [5]. H. Deng discussed fuzzy multi-criteria decision maker approach with pair-wise comparison of weight [6]. H. Fazlollahtabare et al. extended the fuzzy AHP to analyse the alternatives decision [7]. Hwang and Yoon explained TOPSIS method for taking effective decision by ranking the alternatives [8]. Z. Gungor et al. studied personnel selection problem based on different criteria using fuzzy AHP technique [9]. W. Ho described detail literature review along with overall procedure of analytic hierarchy process (AHP) and its applications [10]. A. A. Jessop applied minimally biased weight technique for staff selection [11]. Kahraman and Tolga investigated an alternative ranking method in multi-criteria decision making [12]. Kong and Liu determined the drawbacks and opportunities for success of e-commerce [13]. Kubler et al. established the detail survey in chronological order of fuzzy analytic hierarchy process (FAHP) and its applications [14]. Leung and Cao considered tolerance deviation for the alternatives in the fuzzy AHP [15]. H. Lin examined the similarities and differences between high and low experience groups for evaluation of course website quality [16]. Mikhailov and Tsvetinov considered the linguistic variables for pair-wise comparison

Selection of Hospital Using Integrated Fuzzy AHP and Fuzzy TOPSIS Method 73

of each criteria to develop the matrix with fuzziness for priorities and fuzzy AHP approach is applied for evaluation of services [17–19]. S. P. Ng et al. thoroughly revise and review the risk assessment, management and risk evaluation using FAHP with the classification has been studied [20]. R. Ravangrad et al. described a model in which fuzzy AHP is used to analyse bed management in a hospital [21]. N. Shahbod et al. used AHP to identify and fixed the priority for performance indicators to enhance the hospital facility [22]. Z. Stevic et al. used a combination of FAHP and TOPSIS to evaluate the supplier [23]. Somayeh Nazari et al. developed an expert system based on FAHP and fuzzy inference system in order to obtain the condition of patients who are being examined for heart diseases [24]. C. C. Sun evaluated the performance of the model using fuzzy AHP and TOPSIS method [25]. Vaidya and Kumar gave a brief overview of AHP and its usability [26]. Velasquez and Hester analysed the multi-criteria decision-making methods for its better performance by taking the variations in existing method [27]. Wang and Chen used fuzzy linguistic preference relations for improving consistency in the fuzzy AHP technique [28].

In this chapter, FAHP and fuzzy TOPSIS technique are employed to carry out the selection of the best hospital. FAHP and fuzzy TOPSIS are the most popular methods to solve the decision-making problems and a well-known part of MCDM. MCDM is well known and capable of judging the different domain problems as per the considered attributes.

This chapter is arranged in following manner. Section 5.2 explains the preliminaries of fuzzy sets and fuzzy numbers. Saaty's AHP is discussed in Section 5.3. In Section 5.4, fuzzy AHP is explained with Chang's extent approach. The TOPSIS method is described in Section 5.5. Selection criteria for hospital selection are mentioned in Section 5.6. The evaluation framework for fuzzy AHP and TOPSIS is given in Section 5.7. The outline is concluded in Section 5.8.

5.2 PRELIMINARIES OF FUZZY SETS AND FUZZY NUMBERS

Zadeh (1965) introduced the fuzzy set theory (FST) to describe the ambiguity [29]. FST is capable of dealing with higher-order uncertainty. Mathematical operations and programming are performed in the fuzzy domain. The fuzzy set is a class of objects with rank which is known as membership function. This membership function is classified between 0 and 1.

A triangular fuzzy number (TFN) m is depicted in Figure 5.1. A TFN is represented by (l,m,u). The parameters l, m and u $(l \leq m \leq u)$ indicate the least value, most accurate value and highest value, respectively, to elaborate a fuzzy event.

Membership function of TFN is described in linear fashion as follows:

$$\mu\left(\frac{x}{M}\right) = \begin{cases} 0, & x < l \\ \dfrac{x-l}{m-l}, & l \leq x \leq m \\ \dfrac{u-x}{u-m}, & m \leq x \leq u \\ 0, & x > u \end{cases}$$

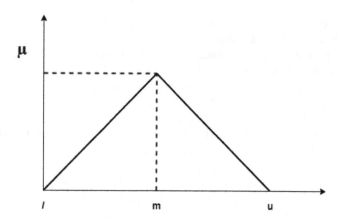

FIGURE 5.1 Membership function of TFN.

A fuzzy number (FN) is represented by left and right degree of membership, described by $\tilde{M}$:

$$\tilde{M} = M^{l(y)}M^{r(y)} = [l+(m-l)y, u+(m-u)y], y\varepsilon[0,1]$$

The left- and the right-side fuzzy number indicate by $l(y)$ and $r(y)$, respectively. The AHP and TOPSIS process is applied to obtain ranking results. However, several operations on TFNs are illustrated in this chapter.

Two positive TFNs $(\zeta_1, \zeta_2, \zeta_3)$ and (η_1, η_2, η_3) have been given as follows:

$$(\zeta_1, \zeta_2, \zeta_3) + (\eta_1, \eta_2, \eta_3) = (\zeta_1 + \eta_1, \zeta_2 + \eta_2, \zeta_3 + \eta_3)$$

$$(\zeta_1, \zeta_2, \zeta_3) - (\eta_1, \eta_2, \eta_3) = (\zeta_1 - \eta_1, \zeta_2 - \eta_2, \zeta_3 - \eta_3)$$

$$(\zeta_1, \zeta_2, \zeta_3) \times (\eta_1, \eta_2, \eta_3) = (\zeta_1 \times \eta_1, \zeta_2 \times \eta_2, \zeta_3 \times \eta_3)$$

$$(\zeta_1, \zeta_2, \zeta_3)/(\eta_1, \eta_2, \eta_3) = (\zeta_1/\eta_1, \zeta_2/\eta_2, \zeta_3/\eta_3)$$

5.3 ANALYTIC HIERARCHY PROCESS (AHP)

Saaty (1980) proposed the analytic hierarchy process (AHP) technique, which is a well-known multi-criteria decision-making (MCDM) tool [30]. The traditional AHP technique is not capable of expressing the correct judgment of the decision makers because of unbalanced scale of decision and its incapability to address in it ambiguity and imprecision in the couple of contrast processes.

The AHP procedure includes seven essential phases, as shown in Figure 5.2.

There are three main components of AHP hierarchy: construction, priority analysis and consistency verification. According to the first component, judges makers will have to down fall their problem with the arrangement of attributes into multiple hierarchical order. The judges have to examine the problem in a pair-wise manner

Selection of Hospital Using Integrated Fuzzy AHP and Fuzzy TOPSIS Method

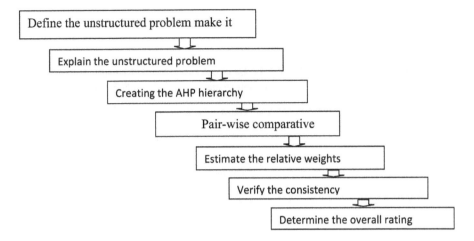

FIGURE 5.2 Structure of AHP procedure.

based on their weight. Sometimes the inconsistency may occur when two attributes have the same criteria which deviate the judgment. For obtaining the consistent judgment, we have to include the degree of compatibility among the couple of comparative studies by the compatibility ratio. If the judges do not obtain the desired level of consistency, then the process has to review and revise. A couple of contrasts are obtained at each step and found to be consistent. Subsequently the judgment for synthesis has to decide the priority ranking of each criterion and its attributes.

5.4 FUZZY ANALYTIC HIERARCHY PROCESS (FAHP)

The AHP construct design for decisions that use a one-dimension hierarchical approach in respect of the judgment steps. The hierarchy is formed in the middle phase, with other judgment at the backside, as shown in Figure 5.2. The FAHP technique gives the design for tuning priorities on every level of the hierarchy applying a couple of contrasts which are quantified and applying a 1 to 9 scale as shown in Figure 5.3 and Table 5.1.

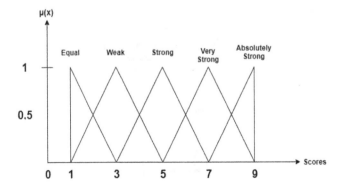

FIGURE 5.3 Linguistic variables describing weights of criteria.

TABLE 5.1
Triangular Fuzzy Conversion Scale

Triangular Fuzzy Number Scale	The Relative Importance of the Two Sub-Elements	Triangular Fuzzy Number Scale
$(1,1,3)$	Equal importance	$\left(\dfrac{1}{3},1,1\right)$
$(1,3,5)$	Slight importance	$\left(\dfrac{1}{5},\dfrac{1}{3},1\right)$
$(3,5,7)$	Important	$\left(\dfrac{1}{7},\dfrac{1}{5},\dfrac{1}{3}\right)$
$(5,7,9)$	Strong importance	$\left(\dfrac{1}{9},\dfrac{1}{7},\dfrac{1}{5}\right)$
$(7,9,9)$	Very strong importance	$\left(\dfrac{1}{9},\dfrac{1}{9},\dfrac{1}{7}\right)$

5.4.1 CHANG'S EXTENT ANALYSIS OF FAHP

Chang proposed a novel approach to take control of fuzzy AHP by using a TFN pair-wise comparison scale of fuzzy AHP and applying the extent analysis method for artificial extent values of the couple of contrasts. Let's assume that $X = \{x_1, x_2, x_3, \ldots\ldots\ldots, x_n\}$ is number of objects and $U = \{u_1, u_2, u_3, \ldots\ldots\ldots, u_n\}$ is the number of goals (a_i). For every object, (a_i) is perform one by one according to the analysis of Chang. m expresses the value of expanded analysis for every object.

$$N_{ai}^1, N_{ai}^2, N_{ai}^3, \ldots\ldots\ldots, N_{ai}^m, \ i = 1, 2, 3, \ldots, n$$

where $N_a^j, \ j = 1, 2, 3, \ldots., m$ are fuzzy triangular numbers.

Extended analysis is proposed by Chang in the following steps:

Step 1: Fuzzy extension for the *ith* object are expressed as

$$P_i = \sum_{j=1}^{n} N_{ai}^j \times \left[\sum_{i=1}^{n} \sum_{j=1}^{m} N_{ai}^j \right]^{-1} \tag{5.1}$$

In order to obtain expression

$$\left[\sum_{i=1}^{n} \sum_{j=1}^{m} N_{ai}^j \right]^{-1}$$

Selection of Hospital Using Integrated Fuzzy AHP and Fuzzy TOPSIS Method

To perform additional fuzzy operations for m are essential for extending analysis, expressed as follows:

$$\sum_{j=1}^{n} N_{ai}^{j} = \left(\sum_{j=1}^{m} l_j, \sum_{j=1}^{m} m_j, \sum_{j=1}^{m} u_j \right)$$

$$\sum_{i=1}^{n}\sum_{j=1}^{m} N_{ai}^{j} = \left(\sum_{i=1}^{n} l_i, \sum_{i=1}^{n} m_i, \sum_{i=1}^{n} u_i \right)$$

Evaluation of the inverse vector is requisite:

$$\left[\sum_{i=1}^{n}\sum_{j=1}^{m} N_{ai}^{j} \right]^{-1} = \left(\frac{1}{\sum_{i=1}^{n} l_i}, \frac{1}{\sum_{i=1}^{n} m_i}, \frac{1}{\sum_{i=1}^{n} u_i} \right) \quad (5.2)$$

Step 2: Possible degree $P_2 \geq P_1$ is defined as

$$V(P_2 \geq P_1) = \begin{cases} 1, & \text{if } m_2 \geq m_1 \\ 0, & \text{if } l_1 \geq u_2 \\ \dfrac{l_1 - u_2}{(m_2 - u_2) - (m_1 - l_1)} & \text{otherwise} \end{cases} \quad (5.3)$$

where d is the largest cross section between μP_1 and μP_2 ordinate as given in Figure 5.4

Comparison of P_1 and P_2 is required as $V(P_1 \geq P_2) \, i \, V(P_2 \geq P_1)$.

Step 3: Possible convex fuzzy number should be larger than k convex number $P_i(i = 1,2,3,\ldots\ldots,k)$ can be explained by

$$V(P_i \geq P_1, P_2, \ldots\ldots, P_k) = \min V(P_i \geq P_k) = w'(P_i) \quad (5.4)$$

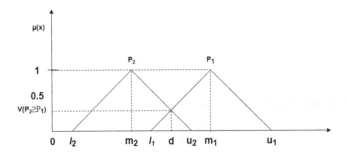

FIGURE 5.4 Common parts of P_2 and P_1.

$$d'(\Psi_i) = \min V(P_i \geq P_k), \ k \neq i, \ k = 1,2, \ldots\ldots\ldots, n$$

The weight vector is mentioned as

$$W_t' = \left(d'(\Psi_1), d'(\Psi_2), \ldots\ldots\ldots, d'(\Psi_n)\right)^T \tag{5.5}$$

where $\Psi_i (i = 1, 2, \ldots\ldots\ldots, n)$ are n elements.

Step 4: The weight vector is diminished by normalisation:

$$W_t = \left(d(\Psi_1), d(\Psi_2), \ldots\ldots\ldots, d(\Psi_n)\right) \tag{5.6}$$

where W_t is a non-fuzzy number.

5.5 TOPSIS METHOD

TOPSIS was first proposed by Hwang and Yoon in 1981 [8]. The TOPSIS method provides us with the ability to rank alternatives based on relative similarity for ideal solution. Ideal solution is classified in two ways: positive and negative. Positive ideal solutions give maximum benefit criteria and minimum cost criteria. Negative ideal solutions emphasise minimum benefit criteria and maximum cost criteria. If the similar indexing for positive and negative ideal solution are achieved, then this situation has to be avoided.

We formulate the TOPSIS technique in following steps:

Initial matrix:

$$X = \|x_{ij}\|_{mn} \tag{5.7}$$

Step 1: Normalise the initial matrix:

$$\|X\| \rightarrow \|R\| \tag{5.8}$$

$$R = \|r_{ij}\|_{mn} \tag{5.9}$$

$$r_{ij} = \frac{x_{ij}}{\sqrt{\sum_{i=1}^{m} x_{ij}^{2}}} \tag{5.10}$$

Step 2: Establish normalised matrix for weights:

$$\|X\| \rightarrow \|R\| \tag{5.11}$$

$$V = \|v_{ij}\| = \|W' r_{ij}\| \tag{5.12}$$

Step 3: Develop the $+ve$ ideal and $-ve$ ideal solution:
D^+ indicates $+ve$ ideal solution and has the most superior characteristics for all criteria:

$$D^+ = \left\{\left(\max_i \vartheta_{ij} \mid j \in E'\right) i \left(\min_i \vartheta_{ij} \mid j \in E''\right)\right\} = \left\{\vartheta_1^+, \vartheta_2^+, \vartheta_3^+ \dots \vartheta_j^+, \dots \vartheta_n^+\right\}, i = \overline{1,m}$$

$$(5.13)$$

$E' \subseteq E$ where E consists of *min* type criteria.
D^-- Indicates $-ve$ ideal solution and has the most superior characteristics for all criteria:

$$D^- = \left\{\left(\max_i \vartheta_{ij} \mid j \in E'\right) i \left(\min_i v_{ij} \mid j \in E'\right)\right\} = \left\{\vartheta_1^-, \vartheta_2^-, \vartheta_3^- \dots \vartheta_j^-, \dots \vartheta_n^-\right\}, i = \overline{1,m}$$

$$(5.14)$$

$E' \subseteq E$ where E consists of *min* type criteria.
Step 4: Calculate the length of each substitute of the PIS (positive ideal solution) and NIS (negative ideal solution):
The distance D^+ can be calculate as:

$$D_i^+ = \sum_{j=1}^{n} d\left(\vartheta_{ij}, \vartheta_j^+\right), i = 1,2,3\dots, m \qquad (5.15)$$

The distance D^- can be calculate as:

$$D_i^- = \sum_{j=1}^{n} d\left(\vartheta_{ij}, \vartheta_j^-\right), i = 1,2,3\dots, m \qquad (5.16)$$

where distance between two fuzzy numbers $\alpha = (a_1, a_2, a_3)$ and $\beta = (b_1, b_2, b_3)$ can be calculated as:

$$d_v(\alpha, \beta) = \sqrt{\frac{1}{3}\left[(a_1 - b_1)^2 + (a_2 - b_2)^2 + (a_3 - b_3)^2\right]} \qquad (5.17)$$

Step 5: Calculate the relative closeness coefficient.
The nearness coefficient CC_i is explained to find the rank of all alternatives. The index CC_i representing the alternative is closed to the *FPIS* (D_i^+) and distance from the *FNIS* (D_i^-). The nearness coefficient of each evaluation Hospital quality can be calculated as:

$$CC_i = \frac{D_i^+}{D_i^+ + D_i^-} \qquad (5.18)$$

Step 6: Ranking of alternatives:
Arrange CC_i values according to their rank, systematically in decreasing order.

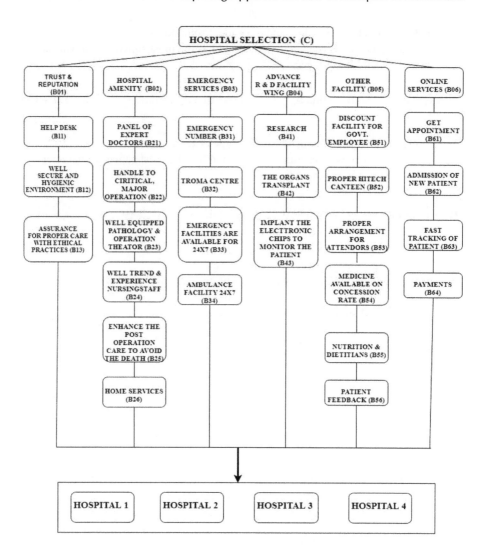

FIGURE 5.5 Proposed model for selection of hospital.

5.6 SELECTION CRITERIA OF HOSPITAL

Selection of the best hospital selection with suitable treatment is a challenging task. Here we emphasise how to identify the best hospital on the basis of the considered criteria and sub-criteria. The broad spectrum of hospital selection is comprehensively evaluated in Figure 5.5.

5.7 EVALUATION FRAMEWORK

In this section, we develop the comparison matrix, based on criteria and sub-criteria by using the fuzzy AHP and TOPSIS method. In this study, we consider all the

Selection of Hospital Using Integrated Fuzzy AHP and Fuzzy TOPSIS Method **81**

criteria and sub-criteria and their relative decision variables. This process is very impactful to make the final decision. The criteria and sub-criteria are prepared by literature and experts' opinions.

5.7.1 Implication of Fuzzy AHP

First, we examine the problem by using fuzzy AHP. We prepare the questionnaire forms for judges and developed pair-wise comparison against other importance. Judges apply the linguistic expression to calculate the ranking of alternatives and converted it into TFN. In TFN, 1–9 scale of fuzzy linguistic variables are expressed for a couple of contrasts. Weights of criteria and sub-criteria are evaluated, and further sub-criteria comprehensive weight are multiplied by the corresponding main criteria weights to get final weight of the sub criteria results are discussed (see Tables 5.2 and 5.3).

TABLE 5.2

Fuzzy Comparison Matrix between Criteria and Sub-Criteria

		B01	B02	B03	B04	B05	B06
B01	B11	$(1, 1, 1)$	$(3, 5, 7)$	$\left(\frac{1}{7}, \frac{1}{5}, \frac{1}{3}\right)$	$(3, 5, 7)$	$\left(\frac{1}{7}, \frac{1}{5}, \frac{1}{3}\right)$	$\left(\frac{1}{7}, \frac{1}{5}, \frac{1}{3}\right)$
	B12	$(1, 1, 1)$	$\left(\frac{1}{9}, \frac{1}{9}, \frac{1}{7}\right)$	$(7, 9, 9)$	$(7, 9, 9)$	$(7, 9, 9)$	$\left(\frac{1}{9}, \frac{1}{9}, \frac{1}{7}\right)$
	B13	$(1, 1, 1)$	$(7, 9, 9)$	$\left(\frac{1}{9}, \frac{1}{9}, \frac{1}{7}\right)$	$(7, 9, 9)$	$(7, 9, 9)$	$\left(\frac{1}{9}, \frac{1}{9}, \frac{1}{7}\right)$
B02	B21	$\left(\frac{1}{9}, \frac{1}{9}, \frac{1}{7}\right)$	$(1, 1, 1),$	$\left(\frac{1}{9}, \frac{1}{9}, \frac{1}{7}\right)$	$(7, 9, 9)$	$\left(\frac{1}{9}, \frac{1}{9}, \frac{1}{7}\right)$	$\left(\frac{1}{9}, \frac{1}{9}, \frac{1}{7}\right)$
	B22	$(5, 7, 9\ )$	$(1, 1, 1)$	$(5, 7, 9)$	$(5, 7, 9)$	$\left(\frac{1}{9}, \frac{1}{7}, \frac{1}{5}\right)$	$\left(\frac{1}{9}, \frac{1}{7}, \frac{1}{5}\right)$
	B23	$\left(\frac{1}{5}, \frac{1}{3}, 1\right)$	$(1, 1, 1)$	$\left(\frac{1}{5}, \frac{1}{3}, 1\right)$	$(1, 3, 5)$	$\left(\frac{1}{5}, \frac{1}{3}, 1\right)$	$\left(\frac{1}{5}, \frac{1}{3}, 1\right)$
	B24	$(7, 7, 9)$	$(1, 1, 1)$	$\left(\frac{1}{9}, \frac{1}{7}, \frac{1}{7}\right)$	$(7, 7, 9)$	$\left(\frac{1}{9}, \frac{1}{7}, \frac{1}{7}\right)$	$\left(\frac{1}{9}, \frac{1}{7}, \frac{1}{7}\right)$
	B25	$\left(\frac{1}{7}, \frac{1}{5}, \frac{1}{3}\right)$	$(1, 1, 1)$	$\left(\frac{1}{7}, \frac{1}{5}, \frac{1}{3}\right)$	$(3, 5, 7)$	$\left(\frac{1}{7}, \frac{1}{5}, \frac{1}{3}\right)$	$\left(\frac{1}{7}, \frac{1}{5}, \frac{1}{3}\right)$
	B26	$(1, 3, 5)$	$(1, 1, 1)$	$\left(\frac{1}{5}, \frac{1}{3}, 1\right)$	$(1, 3, 5)$	$\left(\frac{1}{5}, \frac{1}{3}, 1\right)$	$\left(\frac{1}{5}, \frac{1}{3}, 1\right)$
B03	B31	$(1, 1, 1)$	$\left(\frac{1}{5}, \frac{1}{3}, 1\right)$	$(1, 1, 1)$	$(1, 3, 5)$	$\left(\frac{1}{5}, \frac{1}{3}, 1\right)$	$(1, 3, 5)$
	B32	$(3, 5, 7)$	$\left(\frac{1}{7}, \frac{1}{5}, \frac{1}{3}\right)$	$(1, 1, 1)$	$(3, 5, 7)$	$\left(\frac{1}{7}, \frac{1}{5}, \frac{1}{3}\right)$	$\left(\frac{1}{7}, \frac{1}{5}, \frac{1}{3}\right)$
	B33	$(1, 1, 1)$	$\left(\frac{1}{9}, \frac{1}{7}, \frac{1}{5}\right)$	$(1, 1, 1)$	$(5, 7, 9)$	$\left(\frac{1}{9}, \frac{1}{7}, \frac{1}{5}\right)$	$\left(\frac{1}{9}, \frac{1}{7}, \frac{1}{5}\right)$
	B34	$(1, 1, 3)$	$\left(\frac{1}{3}, 1, 1\right)$	$(1, 1, 1)$	$(1, 1, 3)$	$(1/3, 1, 1)$	$(1, 1, 3)$

(continued)

TABLE 5.2 (*Continued*)
Fuzzy Comparison Matrix between Criteria and Sub-Criteria

		B01	B02	B03	B04	B05	B06
B04	B41	$\left(\frac{1}{3},1,1\right)$	$\left(\frac{1}{3},1,1\right)$	$(1,1,3)$	$(1,1,1)$	$\left(\frac{1}{3},1,1\right)$	$(1,1,3)$
	B42	$(1,3,5)$	$(1,3,5)$	$(1,3,5)$	$(1,1,1)$	$\left(\frac{1}{5},\frac{1}{3},1\right)$	$\left(\frac{1}{5},\frac{1}{3},1\right)$
	B43	$(3,5,7)$	$\left(\frac{1}{7},\frac{1}{5},\frac{1}{3}\right)$	$(3,5,7)$	$(1,1,1)$	$\left(\frac{1}{7},\frac{1}{5},\frac{1}{3}\right)$	$(3,5,7)$
B05	B51	$\left(\frac{1}{9},\frac{1}{7},\frac{1}{7}\right)$	$\left(\frac{1}{9},\frac{1}{7},\frac{1}{7}\right)$	$(3,5,7)$	$(7,7,9)$	$(1,1,1)$	$\left(\frac{1}{9},\frac{1}{7},\frac{1}{7}\right)$
	B52	$(5,7,9)$	$\left(\frac{1}{9},\frac{1}{7},\frac{1}{5}\right)$	$\left(\frac{1}{9},\frac{1}{9},\frac{1}{7}\right)$	$(5,7,9)$	$(1,1,1)$	$\left(\frac{1}{9},\frac{1}{7},\frac{1}{5}\right)$
	B53	$\left(\frac{1}{9},\frac{1}{9},\frac{1}{7}\right)$	$\left(\frac{1}{9},\frac{1}{9},\frac{1}{7}\right)$	$\left(\frac{1}{7},\frac{1}{5},\frac{1}{3}\right)$	$(7,9,9)$	$(1,1,1)$	$\left(\frac{1}{9},\frac{1}{9},\frac{1}{7}\right)$
	B54	$(3,5,7)$	$\left(\frac{1}{7},\frac{1}{5},\frac{1}{3}\right)$	$\left(\frac{1}{5},\frac{1}{3},1\right)$	$(3,5,7)$	$(1,1,1)$	$\left(\frac{1}{7},\frac{1}{5},\frac{1}{3}\right)$
	B55	$\left(\frac{1}{5},\frac{1}{3},1\right)$	$\left(\frac{1}{5},\frac{1}{3},1\right)$	$(1/3,1,1)$	$(1,3,5)$	$(1,1,1)$	$\left(\frac{1}{5},\frac{1}{3},1\right)$
	B56	$(1,1,3)$	$\left(\frac{1}{3},1,1\right)$	$\left(\frac{1}{9},\frac{1}{9},\frac{1}{7}\right)$	$(1,1,3)$	$(1,1,1)$	$\left(\frac{1}{3},1,1\right)$
B06	B61	$\left(\frac{1}{9},\frac{1}{9},\frac{1}{7}\right)$	$\left(\frac{1}{9},\frac{1}{9},\frac{1}{7}\right)$	$(5,7,9)$	$(7,9,9)$	$\left(\frac{1}{9},\frac{1}{9},\frac{1}{7}\right)$	$(1,1,1)$
	B62	$(5,7,9)$	$\left(\frac{1}{9},\frac{1}{7},\frac{1}{5}\right)$	$\left(\frac{1}{9},\frac{1}{7},\frac{1}{5}\right)$	$(5,7,9)$	$\left(\frac{1}{9},\frac{1}{7},\frac{1}{5}\right)$	$(1,1,1)$
	B63	$\left(\frac{1}{9},\frac{1}{7},\frac{1}{5}\right)$	$\left(\frac{1}{9},\frac{1}{7},\frac{1}{5}\right)$	$\left(\frac{1}{7},\frac{1}{5},\frac{1}{3}\right)$	$(5,7,9)$	$\left(\frac{1}{9},\frac{1}{7},\frac{1}{5}\right)$	$(1,1,1)$
	B64	$(3,5,7)$	$\left(\frac{1}{7},\frac{1}{5},\frac{1}{3}\right)$	$\left(\frac{1}{5},\frac{1}{3},1\right)$	$(3,5,7)$	$\left(\frac{1}{7},\frac{1}{5},\frac{1}{3}\right)$	$(1,1,1)$

To get fuzzy combination for each of the criteria, we evaluate $\Sigma_{j=1}^{n} N_{ai}^{j}$ value for each row of the matrix (see Tables 5.4–5.6):

$$CI = (1+1.326+0.481+5.278+1.913+0.121,1+1.710+0.585+7.399$$
$$+2.530+0.135,1+2.080+0.754+8.277+3+0.189)$$

$$CI = (10.11813.359,15.300)$$

Calculation of $\Sigma_{i=1}^{n} \Sigma_{j=1}^{n} N_{ai}^{j}$ as:

$$\sum_{i=1}^{n}\sum_{j=1}^{n} N_{ai}^{j} = (10.118,13.359,15.300)+(5.245,7.970,11.026)$$
$$+(6.455,9.602,12.770)+(4.418,7.981,13.126)+(4.4439,7.119,10.071)$$
$$+(6.986,9.394,11.569)=(37.661,55.423,73.863)$$

TABLE 5.3

Fuzzy Weight Matrix by Taking Geometric Mean

	B01	B02	B03	B04	B05	B06
B01	(1, 1, 1)	(1.326, 1.710, 2.080)	(0.481, 0.585, 0.754)	(5.278, 7.399, 8.277)	(1.913, 2.530, 3.000)	(0.121, 0.135, 0.189)
B02	(0.693, 1.014, 1.638)	(1, 1, 1)	(0.266, 0.368, 0.628)	(3.004, 5.203, 7.095)	(0.141, 0.192, 0.333)	(0.141, 0.192, 0.333)
B03	(1.316, 1.495, 2.141)	(0.180, 0.312, 0.508)	(1, 1, 1)	(3.201, 5.544, 7.297)	(0.180, 0.312, 0.508)	(0.577, 0.937, 1.316)
B04	(1.000, 2.466, 3.271)	(0.362, 0.843, 1.186)	(1, 2.080, 4.217)	(1, 1, 1)	(0.212, 0.405, 0.693)	(0.843, 1.186, 2.759)
B05	(0.577, 0.755, 1.252)	(0.153, 0.231, 0.333)	(0.289, 0.442, 0.628)	(2.265, 4.460, 6.525)	(1, 1, 1)	(0.153, 0.231, 0.333)
B06	(0.656, 0.863, 1.158)	(0.118, 0.146, 0.209)	(0.306, 0.386, 0.541)	(4.787, 6.853, 8.452)	(0.118, 0.146, 0.209)	(1, 1, 1)

TABLE 5.4
Calculation of *CI*

	l	*m*	*u*
B01	10.118	13.359	15.300
B02	5.245	7.970	11.026
B03	6.455	9.602	12.770
B04	4.418	7.981	13.126
B05	4.439	7.119	10.071
B06	6.986	9.394	11.569
$\sum$	**37.661**	**55.423**	**73.863**

$$P_i = \sum_{j=1}^{n} N_{ai}^{j} \times \left[\sum_{i=1}^{n} \sum_{j=1}^{n} N_{ai}^{j} \right]^{-1}$$

$$P_1 = (10.118,\ 13.359, 15.300) \times \left(\frac{1}{37.661}, \frac{1}{55.423}, \frac{1}{73.863} \right)$$

V (preference order values) are calculated as:

$$V(P_l \geq P_2) = \begin{cases} 1, & if\ m_2 \geq m_1 \\ 0, & if\ l_1 \geq u_2 \\ \dfrac{l_1 - u_2}{(m_2 - u_2) - (m_1 - l_1)} & otherwise \end{cases}$$

Priorities of the weight are calculated using Equation (5.9):

$$d'(B01) = \min(1,1,1,1,1) = 1$$

$$d'(B02) = \min(0.737, 0.998, 1, 0.739) = 0.737$$

$$d'(B03) = \min(0.414, 1, 1, 1, 1) = 0.414$$

$$d'(B04) = \min(0.359, 1, 0.805, 1, 0.813) = 0.359$$

TABLE 5.5
Calculation of P_i

	u	*l*	*m*
P_1	0.137	0.241	0.406
P_2	0.071	0.144	0.293
P_3	0.087	0.173	0.339
P_4	0.060	0.144	0.349
P_5	0.060	0.128	0.267
P_6	0.095	0.169	0.307

TABLE 5.6
Calculation of V (Prioritisation)

		$V(P_2 \geq P_1)$			
$V(P_1 \geq P_2)$ 1	$V(P_2 \geq P_3)$ 0.737	0.414	$V(P_4 \geq P_1)$ 0.359	$V(P_5 \geq P_2)$ 0.850	$V(P_6 \geq P_1)$ 0.268
$V(P_1 \geq P_3)$ 1	$V(P_2 \geq P_4)$ 0.998	$V(P_3 \geq P_2)$ 1	$V(P_4 \geq P_2)$ 1	$V(P_5 \geq P_3)$ 0.593	$V(P_6 \geq P_2)$ 1
$V(P_1 \geq P_4)$ 1	$V(P_2 \geq P_5)$ 1	$V(P_3 \geq P_4)$ 1	$V(P_4 \geq P_3)$ 0.805	$V(P_5 \geq P_4)$ 0.868	$V(P_6 \geq P_3)$ 0.961
$V(P_1 \geq P_5)$ 1	$V(P_2 \geq P_6)$ 0.739	$V(P_3 \geq P_5)$ 1	$V(P_4 \geq P_5)$ 1	$V(P_5 \geq P_6)$ 0.576	$V(P_6 \geq P_4)$ 1
$V(P_1 \geq P_6)$ 1		$V(P_3 \geq P_6)$ 1	$V(P_4 \geq P_6)$ 0.813		$V(P_6 \geq P_5)$ 1

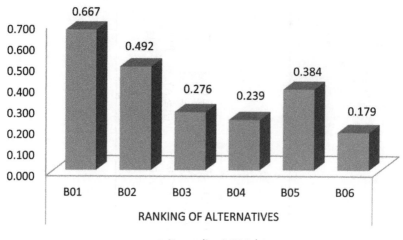

FIGURE 5.6 Weights of criteria using FAHP.

$$d'(B05) = \min(0.850, 0.593, 0.868, 0.576) = 0.576$$

$$d'(B06) = \min(0.268, 1, 0.961, 1, 1) = 0.268$$

The weight vector is shown by

$$W_t' = \left(d'(B01), d'(B02), d'(B03), d'(B04), d'(B05), d'(B06)\right)$$

$$W_t' = (1, 0.737, 0.414, 0.359, 0.576, 0.268)$$

Normalisation of the weight vector is reduced to

$$W_t = (0.667, 0.492, 0.276, 0.239, 0.384, 0.179)$$

The normalised weight of each success factor is shown in Figure 5.6. We identify $B01$ as the best alternative criterion among the other existing criteria. However, $B06$ is the poorest criterion among the other existing criterion.

5.7.2 Implementation of TOPSIS

The TOPSIS method is very applicable and practical for solving real-world problems. It can be also extended according to the requirement of the problem whether theoretical or practical. This depends only on interval or fuzzy weights to model imprecision, uncertainty or vagueness. The TOPSIS method is employed in the following manner:

- Design the fuzzy decision matrix.
- Normalise the fuzzy decision matrix.
- Formulate weighted normalised fuzzy decision matrix.
- Obtain FPIS and FNIS

Selection of Hospital Using Integrated Fuzzy AHP and Fuzzy TOPSIS Method **87**

The TOPSIS method provides the most feasible alternative solution for decision makers. The ranking orders of all alternatives have been obtained (see Tables 5.7–5.12). Table 5.13 shows the final result and rating of H_1, H_2, H_3 and H_4. It is carried out $H_2 > H_4 > H_1 > H_3$.

TABLE 5.7

Initial Matrix for the Compression between Different Hospitals

	H_1	H_2	H_3	H_4
B01	(7,9,9)	(5,7,9)	(3,5,7)	(7,9,9)
B02	(5,7,9)	(3,5,7)	(5,7,9)	(3,5,7)
B03	(3,5,7)	(3,5,7)	(5,7,9)	(1,3,5)
B04	(1,1,3)	(1,3,5)	(1,3,5)	(1,1,3)
B05	(5,7,9)	(3,5,7)	(5,7,9)	(1,3,5)
B06	(3,5,7)	(3,5,7)	(3,5,7)	(1,3,5)
B11	(3,5,7)	(1,13)	(1,3,5)	(1,3,5)
B12	(7,9,9)	(1,3,5)	(3,5,7)	(5,7,9)
B13	(7,9,9)	(7,9,9)	(5,7,9)	(7,9,9)
B21	(7,9,9)	(5,7,9)	(1,3,5)	(1,1,3)
B22	(5,7,9)	(3,5,7)	(7,9,9)	(1,3,5)
B23	(1,3,5)	(3,5,7)	(5,7,9)	(7,9,9)
B24	(7,9,9)	(1,1,3)	(7,9,9)	(5,7,9)
B25	(3,5,7)	(5,7,9)	(1,3,5)	(3,5,7)
B26	(1,3,5)	(7,9,9)	(1,1,3)	(5,7,9)
B31	(1,3,5)	(3,5,7)	(5,7,9)	(1,1,3)
B32	(3,5,7)	(5,7,9)	(7,9,9)	(5,7,9)
B33	(5,7,9)	(1,1,3)	(5,7,9)	(1,3,5)
B34	(1,1,3)	(7,9,9)	(7,9,9)	(3,5,7)
B41	(1,1,3)	(1,3,5)	(3,5,7)	(1,3,5)
B42	(1,3,5)	(3,5,7)	(1,1,3)	(1,1,3)
B43	(3,5,7)	(1,1,3)	(1,3,5)	(3,5,7)
B51	(7,9,9)	(5,7,9)	(5,7,9)	(7,9,9)
B52	(5,7,9)	(3,5,7)	(5,7,9)	(7,9,9)
B53	(7,9,9)	(3,5,7)	(7,9,9)	(1,1,3)
B54	(3,5,7)	(1,1,3)	(3,5,7)	(1,3,5)
B55	(1,3,5)	(1,3,5)	(1,3,5)	(3,5,7)
B56	(1,1,3)	(3,5,7)	(1,1,3)	(5,7,9)
B61	(7,9,9)	(5,7,9)	(3,5,7)	(1,3,5)
B62	(5,7,9)	(3,5,7)	(5,7,9)	(1,1,3)
B63	(5,7,9)	(7,9,9)	(3,5,7)	(5,7,9)
B64	(3,5,7)	(1,3,5)	(1,3,5)	(3,5,7)

TABLE 5.8
Normalised Aggregation Fuzzy Decision Matrix

	H_1	H_2	H_3	H_4
B01	(0.78,1.00,1.00)	(0.56, 0.78, 1.00)	(0.33, 0.56, 0.78)	(0.78, 1.00, 1.00)
B02	(0.56,0.78,1.00)	(0.33, 0.56, 0.78)	(0.56, 0.78, 1.00)	(0.33, 0.56, 0.78)
B03	(0.33,0.56,0.78)	(0.33, 0.56, 0.78)	(0.56, 0.78, 1.00)	(0.11, 0.33, 0.56)
B04	(0.11,0.11,0.33)	(0.11, 0.33, 0.56)	(0.11, 0.33, 0.56)	(0.11, 0.11, 0.33)
B05	(0.56,0.78,1.00)	(0.33, 0.56, 0.78)	(0.56, 0.78, 1.00)	(0.11, 0.33, 0.56)
B06	(0.33,0.56,0.78)	(0.33, 0.56, 0.78)	(0.33, 0.56, 0.78)	(0.11, 0.33, 0.56)
B11	(0.33,0.56,0.78)	(0.11, 0.11, 0.33)	(0.11, 0.33, 0.56)	(0.11, 0.33, 0.56)
B12	(0.78,1.00,1.00)	(0.11, 0.33, 0.56)	(0.33, 0.56, 0.78)	(0.56, 0.78, 1.00)
B13	(0.78,1.00,1.00)	(0.78, 1.00, 1.00)	(0.56, 0.78, 1.00)	(0.78, 1.00, 1.00)
B21	(0.78,1.00,1.00)	(0.56, 0.78, 1.00)	(0.11, 0.33, 0.56)	(0.11, 0.11, 0.33)
B22	(0.56,0.78,1.00)	(0.33, 0.56, 0.78)	(1.00, 1.00, 1.00)	(0.11, 0.33, 0.56)
B23	(0.11,0.33,0.56)	(0.33, 0.56, 0.78)	(0.56, 0.78, 1.00)	(0.78, 1.00, 1.00)
B24	(0.78,1.00,1.00)	(0.11, 0.11, 0.33)	(0.78, 1.00, 1.00)	(0.56, 0.78, 1.00)
B25	(0.33,0.56,0.78)	(0.56, 0.78, 1.00)	(0.11, 0.33, 0.56)	(0.33, 0.56, 0.78)
B26	(0.11,0.33,0.56)	(0.78, 1.00, 1.00)	(0.11, 0.11, 0.33)	(0.56, 0.78, 1.00)
B31	(0.11,0.33,0.56)	(0.33, 0.56, 0.78)	(0.56, 0.78, 1.00)	(0.11, 0.11, 0.33)
B32	(0.33,0.56,0.78)	(0.56, 0.78, 1.00)	(0.78, 1.00, 1.00)	(0.56, 0.78, 1.00)
B33	(0.56,0.78,1.00)	(0.11, 0.11, 0.33)	(0.56, 0.78, 1.00)	(0.11, 0.33, 0.56)
B34	(0.11,0.11,0.33)	(0.78, 1.00, 1.00)	(0.78, 1.00, 1.00)	(0.33, 0.56, 0.78)
B41	(0.11,0.11,0.33)	(0.11, 0.33, 0.56)	(0.33, 0.56, 0.78)	(0.11, 0.33, 0.56)
B42	(0.11,0.33,0.56)	(0.33, 0.56, 0.78)	(0.11, 0.11, 0.33)	(0.11, 0.11, 0.33)
B43	(0.33,0.56,0.78)	(0.11, 0.11, 0.33)	(0.11, 0.33, 0.56)	(0.33, 0.56, 0.78)
B51	(0.78,1.00,1.00)	(0.56, 0.78, 1.00)	(0.56, 0.78, 1.00)	(0.78, 1.00, 1.00)
B52	(0.56,0.78,1.00)	(0.33, 0.56, 0.78)	(0.56, 0.78, 1.00)	(0.56, 0.78, 1.00)
B53	(0.78,1.00,1.00)	(0.33, 0.56, 0.78)	(0.78, 1.00, 1.00)	(0.11, 0.11, 0.33)
B54	(0.33,0.56,0.78)	(0.11, 0.11, 0.33)	(0.33, 0.56, 0.78)	(0.11, 0.33, 0.56)
B55	(0.11,0.33,0.56)	(0.11, 0.33, 0.56)	(0.11, 0.33, 0.56)	(0.33, 0.56, 0.78)
B56	(0.11,0.11,0.33)	(0.33, 0.56, 0.78)	(0.11, 0.11, 0.33)	(0.56, 0.78, 1.00)
B61	(0.78,1.00,1.00)	(0.56, 0.78, 1.00)	(0.33, 0.56, 0.78)	(0.11, 0.33, 0.56)
B62	(0.56,0.78,1.00)	(0.33, 0.56, 0.78)	(0.56, 0.78, 1.00)	(0.11, 0.11, 0.33)
B63	(0.56,0.78,1.00)	(0.78, 1.00, 1.00)	(0.33, 0.56, 0.78)	(0.56, 0.78, 1.00)
B64	(0.33,0.56,0.78)	(0.11, 0.33, 0.56)	(0.11, 0.33, 0.56)	(0.33, 0.56, 0.78)

TABLE 5.9
Normalisation of the Initial Matrix

	H_1	H_2	H_3	H_4
B01	(0.64, 0.59, 0.48)	(0.64, 0.56, 0.52)	(0.31, 0.35, 0.37)	(0.89, 0.78, 0.62)
B02	(0.46, 0.46, 0.48)	(0.38, 0.40, 0.40)	(0.52, 0.49, 0.47)	(0.38, 0.43, 0.48)
B03	(0.28, 0.33, 0.37)	(0.38, 0.40, 0.40)	(0.52, 0.49, 0.47)	(0.13, 0.26, 0.34)
B04	(0.09, 0.07, 0.16)	(0.13, 0.24, 0.29)	(0.10, 0.21, 0.26)	(0.13, 0.09, 0.21)
B05	(0.46, 0.46, 0.48)	(0.38, 0.40, 0.40)	(0.52, 0.49, 0.47)	(0.13, 0.26, 0.34)
B06	(0.28, 0.33, 0.37)	(0.38, 0.40, 0.40)	(0.31, 0.35, 0.37)	(0.13, 0.26, 0.34)
B11	(0.17, 0.33, 0.39)	(0.33, 0.40, 0.47)	(0.41, 0.43, 0.50)	(0.17, 0.11, 0.23)
B12	(0.50, 0.55, 0.55)	(0.55, 0.56, 0.61)	(0.58, 0.56, 0.50)	(0.83, 0.76, 0.70)
B13	(0.83, 0.76, 0.70)	(0.11, 0.08, 0.20)	(0.41, 0.43, 0.50)	(0.17, 0.33, 0.39)
B21	(0.60, 0.56, 0.49)	(0.46, 0.46, 0.48)	(0.08, 0.20, 0.29)	(0.10, 0.07, 0.17)
B22	(0.43, 0.44, 0.49)	(0.28, 0.33, 0.37)	(0.72, 0.59, 0.52)	(0.10, 0.21, 0.28)
B23	(0.09, 0.19, 0.27)	(0.28, 0.33, 0.37)	(0.40, 0.46, 0.52)	(0.67, 0.62, 0.50)
B24	(0.60, 0.56, 0.49)	(0.09, 0.07, 0.16)	(0.56, 0.59, 0.52)	(0.48, 0.48, 0.50)
B25	(0.26, 0.31, 0.38)	(0.46, 0.46, 0.48)	(0.08, 0.20, 0.29)	(0.29, 0.34, 0.39)
B26	(0.09, 0.19, 0.27)	(0.64, 0.59, 0.48)	(0.08, 0.07, 0.17)	(0.48, 0.48, 0.50)
B31	(0.17, 0.33, 0.39)	(0.33, 0.40, 0.47)	(0.41, 0.43, 0.50)	(0.17, 0.11, 0.23)
B32	(0.50, 0.55, 0.55)	(0.55, 0.56, 0.61)	(0.58, 0.56, 0.50)	(0.83, 0.76, 0.70)
B33	(0.83, 0.76, 0.70)	(0.11, 0.08, 0.20)	(0.41, 0.43, 0.50)	(0.17, 0.33, 0.39)
B34	(0.17, 0.11, 0.23)	(0.76, 0.72, 0.61)	(0.58, 0.56, 0.50)	(0.50, 0.55, 0.55)
B41	(0.30, 0.17, 0.33)	(0.30, 0.51, 0.55)	(0.90, 0.85, 0.77)	(0.30, 0.51, 0.55)
B42	(0.30, 0.51, 0.55)	(0.90, 0.85, 0.77)	(0.30, 0.17, 0.33)	(0.30, 0.17, 0.33)
B43	(0.90, 0.85, 0.77)	(0.30, 0.17, 0.33)	(0.30, 0.51, 0.55)	(0.90, 0.85, 0.77)
B51	(0.60, 0.57, 0.50)	(0.68, 0.60, 0.56)	(0.48, 0.48, 0.50)	(0.67, 0.62, 0.50)
B52	(0.43, 0.45, 0.50)	(0.41, 0.43, 0.43)	(0.48, 0.48, 0.50)	(0.48, 0.48, 0.50)
B53	(0.60, 0.57, 0.50)	(0.41, 0.43, 0.43)	(0.67, 0.62, 0.50)	(0.10, 0.07, 0.17)
B54	(0.26, 0.32, 0.39)	(0.14, 0.09, 0.19)	(0.29, 0.34, 0.39)	(0.10, 0.21, 0.28)
B55	(0.09, 0.19, 0.28)	(0.14, 0.26, 0.31)	(0.10, 0.21, 0.28)	(0.29, 0.34, 0.39)
B56	(0.09, 0.06, 0.17)	(0.41, 0.43, 0.43)	(0.10, 0.07, 0.17)	(0.48, 0.48, 0.50)
B61	(0.67, 0.63, 0.53)	(0.55, 0.55, 0.59)	(0.45, 0.48, 0.49)	(0.17, 0.33, 0.39)
B62	(0.48, 0.49, 0.53)	(0.33, 0.39, 0.46)	(0.75, 0.67, 0.63)	(0.17, 0.11, 0.23)
B63	(0.48, 0.49, 0.53)	(0.76, 0.70, 0.59)	(0.45, 0.48, 0.49)	(0.83, 0.76, 0.70)
B64	(0.29, 0.35, 0.41)	(0.11, 0.23, 0.33)	(0.15, 0.29, 0.35)	(0.50, 0.55, 0.55)

TABLE 5.10
Weighted Normalise Fuzzy Matrix

	H_1	H_2	H_3	H_4
B01	(0.05, 0.04, 0.03)	(0.05, 0.04, 0.04)	(0.02, 0.03, 0.03)	(0.07, 0.06, 0.04)
B02	(0.04, 0.03, 0.03)	(0.03, 0.03, 0.03)	(0.04, 0.04, 0.03)	(0.03, 0.03, 0.03)
B03	(0.02, 0.02, 0.03)	(0.03, 0.03, 0.03)	(0.04, 0.04, 0.03)	(0.01, 0.02, 0.02)
B04	(0.01, 0.00, 0.01)	(0.01, 0.02, 0.02)	(0.01, 0.02, 0.02)	(0.01, 0.01, 0.01)
B05	(0.04, 0.03, 0.03)	(0.03, 0.03, 0.03)	(0.04, 0.04, 0.03)	(0.01, 0.02, 0.02)
B06	(0.02, 0.02, 0.03)	(0.03, 0.03, 0.03)	(0.02, 0.03, 0.03)	(0.01, 0.02, 0.02)
B11	(0.01, 0.02, 0.03)	(0.03, 0.03, 0.03)	(0.03, 0.03, 0.04)	(0.01, 0.01, 0.02)
B12	(0.04, 0.04, 0.04)	(0.04, 0.04, 0.04)	(0.04, 0.04, 0.04)	(0.07, 0.06, 0.05)
B13	(0.06, 0.06, 0.05)	(0.01, 0.01, 0.01)	(0.03, 0.03, 0.04)	(0.01, 0.03, 0.03)
B21	(0.05, 0.04, 0.03)	(0.04, 0.04, 0.03)	(0.01, 0.01, 0.02)	(0.01, 0.01, 0.01)
B22	(0.03, 0.03, 0.03)	(0.02, 0.03, 0.03)	(0.06, 0.04, 0.04)	(0.01, 0.02, 0.02)
B23	(0.01, 0.01, 0.02)	(0.02, 0.03, 0.03)	(0.03, 0.03, 0.04)	(0.06, 0.05, 0.04)
B24	(0.05, 0.04, 0.03)	(0.01, 0.01, 0.01)	(0.04, 0.04, 0.04)	(0.04, 0.04, 0.04)
B25	(0.02, 0.02, 0.03)	(0.04, 0.04, 0.03)	(0.01, 0.01, 0.02)	(0.02, 0.03, 0.03)
B26	(0.01, 0.01, 0.02)	(0.05, 0.05, 0.03)	(0.01, 0.01, 0.01)	(0.04, 0.04, 0.04)
B31	(0.01, 0.02, 0.03)	(0.03, 0.03, 0.03)	(0.03, 0.03, 0.04)	(0.01, 0.01, 0.02)
B32	(0.04, 0.04, 0.04)	(0.04, 0.04, 0.04)	(0.04, 0.04, 0.04)	(0.01, 0.03, 0.03)
B33	(0.06, 0.06, 0.05)	(0.01, 0.01, 0.01)	(0.03, 0.03, 0.04)	(0.04, 0.04, 0.04)
B34	(0.01, 0.01, 0.02)	(0.06, 0.06, 0.04)	(0.04, 0.04, 0.04)	(0.03, 0.04, 0.04)
B41	(0.02, 0.01, 0.02)	(0.02, 0.04, 0.04)	(0.07, 0.06, 0.05)	(0.03, 0.01, 0.02)
B42	(0.02, 0.04, 0.04)	(0.07, 0.06, 0.06)	(0.02, 0.01, 0.02)	(0.08, 0.07, 0.06)
B43	(0.07, 0.06, 0.05)	(0.02, 0.01, 0.02)	(0.02, 0.04, 0.04)	(0.06, 0.05, 0.04)
B51	(0.05, 0.04, 0.04)	(0.05, 0.05, 0.04)	(0.04, 0.04, 0.04)	(0.04, 0.04, 0.04)
B52	(0.03, 0.03, 0.04)	(0.03, 0.03, 0.03)	(0.04, 0.04, 0.04)	(0.01, 0.01, 0.01)
B53	(0.05, 0.04, 0.04)	(0.03, 0.03, 0.03)	(0.05, 0.05, 0.04)	(0.01, 0.02, 0.02)
B54	(0.05, 0.04, 0.03)	(0.01, 0.01, 0.01)	(0.02, 0.03, 0.03)	(0.07, 0.06, 0.04)
B55	(0.04, 0.03, 0.03)	(0.05, 0.04, 0.04)	(0.02, 0.03, 0.03)	(0.03, 0.03, 0.03)
B56	(0.02, 0.02, 0.03)	(0.03, 0.03, 0.03)	(0.04, 0.04, 0.03)	(0.01, 0.02, 0.02)
B61	(0.01, 0.00, 0.01)	(0.03, 0.03, 0.03)	(0.04, 0.04, 0.03)	(0.01, 0.01, 0.01)
B62	(0.04, 0.03, 0.03)	(0.01, 0.02, 0.02)	(0.01, 0.02, 0.02)	(0.01, 0.02, 0.02)
B63	(0.02, 0.02, 0.03)	(0.03, 0.03, 0.03)	(0.04, 0.04, 0.03)	(0.01, 0.02, 0.02)
B64	(0.01, 0.02, 0.03)	(0.03, 0.03, 0.03)	(0.02, 0.03, 0.03)	(0.01, 0.01, 0.02)

TABLE 5.11
D^+ (FPIS Distance)

	H1	H2	H3	H4
B01	0.0584	0.0608	0.0707	0.0563
B02	0.0642	0.0693	0.0629	0.0696
B03	0.0718	0.0693	0.0629	0.0818
B04	0.0836	0.0805	0.0802	0.0902
B05	0.0642	0.0693	0.0629	0.0818
B06	0.0612	0.0617	0.0602	0.0566
B11	0.0545	0.0883	0.0660	0.0780
B12	0.0594	0.0658	0.0812	0.0918
B13	0.0653	0.0738	0.0581	0.0850
B21	0.0813	0.0738	0.0653	0.0601
B22	0.0594	0.0894	0.0595	0.0664
B23	0.0728	0.0658	0.0812	0.0749
B24	0.0813	0.0598	0.0873	0.0664
B25	0.0740	0.0701	0.0660	0.0883
B26	0.0612	0.0617	0.0602	0.0566
B31	0.0545	0.0883	0.0660	0.0780
B32	0.0835	0.0565	0.0602	0.0639
B33	0.0784	0.0670	0.0543	0.0686
B34	0.0541	0.0806	0.0656	0.0561
B41	0.0592	0.0592	0.0636	0.0601
B42	0.0651	0.0676	0.0636	0.0664
B43	0.0592	0.0676	0.0579	0.0918
B51	0.0726	0.0875	0.0714	0.0850
B52	0.0812	0.0794	0.0806	0.0749
B53	0.0573	0.0621	0.0639	0.0780
B54	0.0630	0.0705	0.0562	0.0883
B55	0.0669	0.0710	0.0665	0.0736
B56	0.0673	0.0714	0.0663	0.0743
B61	0.0674	0.0715	0.0664	0.0745
B62	0.0672	0.0715	0.0666	0.0742
B63	0.0666	0.0712	0.0660	0.0736
B64	0.0667	0.0713	0.0662	0.0733

TABLE 5.12

D^- (FNIS Distance)

	H1	H2	H3	H4
B01	0.0491	0.0518	0.0272	0.0753
B02	0.0383	0.0335	0.0426	0.0392
B03	0.0237	0.0335	0.0426	0.0198
B04	0.0020	0.0162	0.0126	0.0035
B05	0.0383	0.0335	0.0426	0.0198
B06	0.0237	0.0335	0.0272	0.0198
B11	0.0231	0.0357	0.0389	0.0068
B12	0.0460	0.0538	0.0487	0.0763
B13	0.0460	0.0538	0.0487	0.0763
B21	0.0700	0.0033	0.0389	0.0264
B22	0.0468	0.0416	0.0130	0.0000
B23	0.0369	0.0267	0.0543	0.0135
B24	0.0097	0.0267	0.0409	0.0567
B25	0.0468	0.0000	0.0509	0.0444
B26	0.0229	0.0416	0.0130	0.0285
B31	0.0097	0.0529	0.0005	0.0444
B32	0.0231	0.0357	0.0389	0.0068
B33	0.0460	0.0538	0.0487	0.0763
B34	0.0700	0.0033	0.0389	0.0264
B41	0.0063	0.0669	0.0487	0.0507
B42	0.0159	0.0449	0.0805	0.0456
B43	0.0782	0.0174	0.0435	0.0850
B51	0.0477	0.0568	0.0424	0.0567
B52	0.0377	0.0371	0.0424	0.0444
B53	0.0477	0.0371	0.0537	0.0000
B54	0.0236	0.0031	0.0273	0.0135
B55	0.0102	0.0184	0.0130	0.0285
B56	0.0532	0.0521	0.0418	0.0264
B61	0.0420	0.0344	0.0632	0.0068
B62	0.0420	0.0652	0.0418	0.0763
B63	0.0266	0.0175	0.0212	0.0507
B64	0.0144	0.0133	0.0132	0.0140

TABLE 5.13

Ranking of Alternatives

	H1	H2	H3	H4
CC_i	0.6570	0.6747	0.6389	0.6703
Rank	3	1	4	2

5.8 CONCLUSION

In this chapter, we proposed a general hierarchical model for the selection of best hospitals. Selection under the considerable criteria and sub-criteria is examined by using the fuzzy AHP and TOPSIS technique. First, we applied Chang's extent FAHP with the use of TFN for a couple of contrast scales of fuzzy AHP. After the rigorous evaluation of criteria among the others, it has been observed that criteria $B01$ is the best alternative. After that the TOPSIS method is employed wherein normalised fuzzy decision matrix, weighted normalised fuzzy decision matrix and FIPS and FNIS are determined. In Figure 5.7, it is well identified that H2 is the best alternative solution among the other alternative solutions.

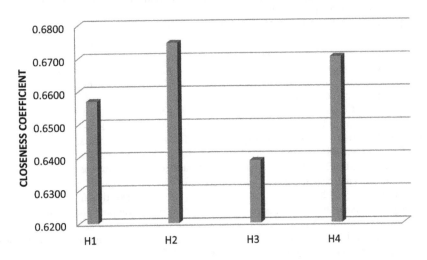

FIGURE 5.7 Ranking of alternatives.

REFERENCES

1. Andres, R., Espinilla, M., and Martinez, L. (2010): An extended hierarchical linguistic model for managing integral evaluation, International Journal of Computational Intelligence System, Vol. 3, pp. 486–500.
2. Bronja, H. (2015): Two phase selection procedure of aluminized sheet supplier by applying fuzzy AHP and fuzzy TOPSIS methodology, Tehnicki Vjesnik, Vol. 22(4), pp. 821–828.
3. Chang, D. Y. (1996): Applications of the extent analysis method on fuzzy AHP, European Journal of Operational Research, Vol. 95, pp. 649–655.
4. Chou, S., Chang, Y., and Shen, C. (2008): A fuzzy simple additive weighting system under group decision-making for facility location selection with objective/subjective attributes, European Journal of Operational Research, Vol. 189(1), pp. 132–145.
5. Dagdevien, M., and Yuksel, I. (2008): Developing a fuzzy analytic hierarchy process (AHP) model for behavior- based safety management, Journal of Information Science, Vol. 178(6), pp. 1717–1733.
6. Deng, H. (1999): Multi-criteria analysis with fuzzy pair-wise comparison, International Journal of Approx Reason, Vol. 21(3), pp. 215–231.

7. Fazlollahtabare, H., Eslami, H., and Salmani, H. (2010): Designing a fuzzy expert system to evaluate alternatives in fuzzy analytic hierarchy process, Journal of Software Engineering and Applications, Vo. 3(4), pp. 409–418.
8. Hwang, C. L., and Yoon, K. (1981): Multiple attributes decision making methods and applications, Springer, Berlin.
9. Gungor, Z., Serhadhoglu, G., and Kesen, S. E. (2009): A fuzzy AHP approach to staff selection problem, Applied Soft Computing, Vol. 9, pp. 641–646.
10. Ho, W. (2008): Integrated analytic hierarchy process and its applications-a literature review, European Journal of Operational Research, Vol. 186(1), pp. 211–228.
11. Jessop, A. (2004): Minimally biased weight determination in staff, Journal of Operation Research, Vol. 153, pp. 433–444.
12. Kahraman, C., Tolga, C. A. (2009): An alternative ranking approach and its usage in multi-criteria decision making, International Journal of Computational Intelligence Systems, Vol. 2, pp. 219–235.
13. Kong, F., and Liu, H. (2005): Applying fuzzy analytic hierarchy process to evaluate success factors of e-commerce, International Journal of Information and Systems Sciences, Vol. 1(3–4), pp. 406–412.
14. Kubler, S., Robert, J., Derigent, J., Voisin, A., and Traon, Y. L. (2016): A state-of-the-art survey and test bed of fuzzy-AHP (FAHP) applications, Expert Systems with Applications, Vol. 65, pp. 398–422.
15. Leung, L. C., and Cao, D. (2000): On consistency and ranking of alternatives in fuzzy AHP, European Journal of Operational Research, Vol. 124, pp.102–113.
16. Lin, H. (2010): An application of fuzzy AHP for evaluating course website quality, Computers and Education, Vol. 54, pp. 877–888.
17. Mikhailov, L. (2002): Fuzzy analytical approach to partnership selection in formation of virtual enterprises, Omega, Vol. 30, pp. 393–401.
18. Mikhailov, L. (2004): A Fuzzy approach to deriving priorities from interval pairwise comparison judgements, European Journal of Operational Research, Vol. 159, pp. 687–704.
19. Mikhailov, L., and Tsvetinov, P. (2004): Evaluation of services using fuzzy analytical Hierarchy process, Applied Soft Computing, Vol. 5, pp. 23–33.
20. Ng, S. P., Ignatius, J., Goh, M., Rahman, A., and Zhan, F. (2017): The state of the art in FAHP in risk assessment, Fuzzy analytic hierarchy process, CRC Press, New York, pp. 11–44.
21. Ravangrad, R., Bahadori, M., Raadabadi, M., Teymourzadeh, E., Alimomohammod Zadeh, K., and Mehrabian, F. (2011): A model for the development of hospital bed using fuzzy AHP, International Journal of Public Health, Vol. 46 (11), pp. 1555–1562.
22. Shahbod, N., Mansouri, N., Bayat, M., Nouri, J., and Ghoddousi, J. (2017): A AHP press to identify and prioritize environment performance indicators in hospitals, International Journal of Occupational Hygiene, Vol. 9, pp. 66–77.
23. Stevic, Z., Tanackov, I., Vasiljevic, M., Novarlic, B., and Stojic, G. (2016): An integrated fuzzy AHP and TOPSIS model for supplier evaluation, Serbian Journal of Management, Vol. 11(1), pp. 15–27.
24. Nazari, Somayeh, Fallah, Mohammad, Kazemipoor, Hamed, and Salehipour, Amir, (2018): A fuzzy inference- fuzzy analytic hierarchy process-based clinical decision support system for diagnosis of heart diseases, Expert Systems with Applications, Vol. 95, pp. 261–271.
25. Sun, C. C. (2010): A performance evaluation model by integrating fuzzy AHP and TOPSIS methods, Expert Systems with Applications, Vol. 37(12), pp. 7745–7754.
26. Vaidya, O.S., and Kumar, S. (2006): Analytic hierarchy process: an overview of applications, European Journal of Operational Research, Vol. 19(1), pp. 1–29.

27. Velasquez, M., and Hester, P. T. (2013): An analysis of multi-criteria decision making methods, International Journal of Operations Research, Vol. 10(2), pp. 56–66.
28. Wang, T. C., and Chen, Y. H. (2008): Appling fuzzy linguistic preference relation to the improvement of consistency of fuzzy AHP, Information Sciences, Vol. 178, pp. 3755–3765.
29. Zadeh, L. (1965): Fuzzy sets information and control, Journal of Information and Control, Vol. 8(3), pp. 338–353.
30. Saaty, T. L. (1980): Multi-criteria decision making the analytic hierarchy process, McGraw-Hill, New York.

6 Computation of Threshold Rate for the Spread of HIV in a Mobile Heterosexual Population and Its Implication for SIR Model in Healthcare

Suresh Rasappan and Regan Murugesan
Department of Mathematics
Vel Tech Rangarajan Dr. Sagunthala
R&D Institute of Science and Technology
Avadi, Chennai, India

CONTENTS

6.1 Introduction 98
6.2 Basic Concepts..99
    6.2.1 Limit Cycle 99
    6.2.2 Bendixson's Criterion 99
    6.2.3 Poincare Bendixons 99
6.3 Basic Reproduction Number (R_0) 100
6.4 The Threshold Rate 100
6.5 Description of the Model... 100
6.6 Analysis of the Behaviour of the HIV–SIR Model 103
6.7 The Spread of HIV within the Population... 103
6.8 The Rate of Removal of HIV Infection... 106
6.9 Analysis of Local Asymptotic Stability 107
6.10 Numerical Simulation... 109
6.11 Conclusion 110
References... 111

6.1 INTRODUCTION

Human immunodeficiency virus (HIV) is one type of virus which damages the immune system of human beings and hence causes a lot of diseases and can reach the stage of acquired immune deficiency syndrome (AIDS) [1]. A person is said to have AIDS when his or her immune system is very infirm to defend oneself against the infection and the person develops certain defining symptoms and illness.

It is one of the uncurable infectious diseases, but there is a possibility to extend the life of an infected person by reducing the viral load from the human body through the antireterovial treatment (ART) [2]. If untreated, serious illness can occur and lead to death. Now in the twenty-first century, AIDS is one of the serious pandemics at the global level.

The primary mode of HIV transmission is unprotected sexual intercourse (includes anal or oral sex), contaminated blood transfusion, hypodermic needles and during pregnancies from mother to child or breast feeding after delivery [3]. The primary way to control the spread of HIV is prevention by individuals themselves through safe sex and the usage of new needles. There is no known cure or vaccines for AIDS so far. However, some social welfare organisations like HAART can reduce or slow down the epidemic of the disease and may lead to normal life expectancy. Since it was first recognised in 1981 by the Centre for Disease Control and Prevention, AIDS has caused over 34 million deaths. As of 2010, approximately 36 million people are living with HIV/AIDS globally and it has caused over 34 million deaths (UNAIDS).

Diseases are major issues in the day-to-day life of our society. They bring a huge epidemic and lead to the loss of human population. Hence formulating a mathematical modelling for the infectious diseases is imperative to control the diseases and to reduce the epidemic [4].

The study of the occurrence of disease is called epidemiology. The transmission of diseases through the individual population is described by the epidemic model [5]. A description of interacting population between the three stages of susceptible, infectible and recovery (SIR) are described in a simple dynamic model in epidemiology. The spread of an infectious disease involves not only the disease-related factors such as the infectious agent, mode of transmission, infectious period, susceptibility, resistance and so on but also social and cultural perspectives.

A basic concept in epidemiology is the existence of threshold. It is a critical value for quantities such as the contact number, population size or vector density that must be exceeded in order for an epidemic to occur or for a disease to remain endemic.

Most of the research works carried out in recent times deal with HIV infections in CD4 cells, T4 helper cells, models on HIV incubation periods and so on. However, in the present work, we mainly focus on the transmission of HIV through the mobile heterosexual population. Nowadays most of the individuals are migrating from one place to another place due to job availability, upgrading their lifestyle, moving closer to family and so forth. Once an HIV-infected individual migrates, the transmission rate of HIV will vary.

This research work is concerned with the threshold rate in mobile heterosexual population dynamics and its implementation with the SIR model. In Section 6.2, the basic concepts are explained. In Section 6.3, the basic reproduction

Computation of Threshold Rate for the Spread of HIV   **99**

number is described. In Section 6.4, the results of the investigation on the SIR model are presented. The behaviour of HIV in SIR model is investigated in Section 6.5. The spread of HIV within the population is analysed in Section 6.6. In Section 6.7, the last-stage HIV infection is analysed. In Section 6.8 conclusions are presented.

6.2 BASIC CONCEPTS

6.2.1 LIMIT CYCLE

While examining a system, one has to find out how trajectories of the system appear in the vicinity of each critical point. A limit cycle is a closed trajectory with a property that at least one other trajectory spirals into it as time t tends to infinity or spiral away from it.

- If the other trajectory spirals into it as t tends to infinity, then it is stable.
- If the limit cycle reaches infinity in the negative side as t tends to infinity, then it is said to be unstable.
- If the limit cycle proceeds towards infinity in the both sides (positive and negative), as t tends to infinity then it is semi-stable.

6.2.2 BENDIXSON'S CRITERION

Bendixson's criterion focuses attention on the nonexistence of limit cycles. In a connected region, if the expression is not identically zero, then there does not exist any periodic orbit in the region.

If u_x and v_y are conditions in a simply connected region R and $\dfrac{\partial u}{\partial x} + \dfrac{\partial v}{\partial y} \neq 0$ at any point of R, then the described family of equations

$$x' = u(x, y)$$
$$y' = v(x, y)$$

does not contain orbits inside R.

6.2.3 POINCARE BENDIXONS

Consider the cluster of differential equation

$$\dot{x} = u(x, y)$$
$$\dot{y} = v(x, y)$$

Assume that $x_{h} = x_{h}(t), y_{h} = y_{h}(t)$ satisfies the above differential equations in a region bounded by the plane without containing its points of equilibrium. Then, its orbit mass spirals into a simply connected curve, which is itself the orbit of the recurring solution of the underlying differential equations.

6.3 BASIC REPRODUCTION NUMBER (R_0)

The main target of mathematical epidemiology is to understand how control and eradicate the infectious disease. The study of autonomous models shows that a disease can cause an epidemic depending on the basic reproduction number (R_0) [6, 7]. It is a measure of indicating how much a disease will be transferred from a single infectious person to another. It is an average number of secondary cases of infections generated from the one by the primary case in a fully susceptible population.

An assumption for the model is that all the individuals are susceptible to HIV infection. The reproduction number R_0 is a measure defined at the entry time of the disease into the population (time = 0). The reproduction number is calculated by the formula

$$R_0 = \beta CD$$

where

- β is the probability of transmission from an HIV-positive individual to an HIV-negative partner,
- C is the rate at which new sexual partnerships are formed per annum and
- D is the average duration in years of HIV infection.

The reproduction number R_0 has the following characteristics:

- If $R_0 > 1$, the transmission or spread of the disease will be raised and there will be an epidemic of it.
- If $R_0 < 1$, the diseases will die out. It means that, with regard to HIV, transmission will be reduced, the infection rate will be decreased and the individual will not reach the stage of AIDS.

6.4 THE THRESHOLD RATE

Threshold refers to the level at which certain event begins to occur. In our study, the concern is the prevention of an affected individual from moving to the next severe state. The threshold rate plays a predominant role in the identification of a safety net in the analysis of different states involved in HIV/AIDS. Consider the scenario in which a small group of humans having infectious HIV is inserted into a huge populace that is capable of catching HIV. A system of differential equations is derived which determines the spread of infectious HIV within a populace, and analysis can be carried out on the behaviour of the solutions [8–11]. This technique results in the threshold theorem of HIV epidemiology, which states that a virus will occur if the number of those individuals who are at risk of the disease exceeds a sure threshold value.

6.5 DESCRIPTION OF THE MODEL

For this study, the populace is divided into three stages of individuals: the susceptible (x_1^r), the infectious (x_2^r) and the removed elegance x_R. The susceptible includes such of those people who are all not infective but are able to catch the HIV and turning

Computation of Threshold Rate for the Spread of HIV

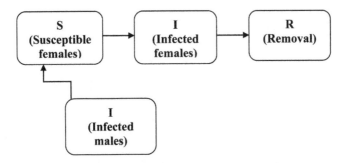

FIGURE 6.1 Compartments of the disease.

into infective [12–14]. The infective elegance is composed of those individuals who are capable of transmitting the HIV to others. The eliminated elegance consists of those people who had the HIV and are lifeless (Figure 6.1).

HIV damages the CD4+ cells and reduces its count in the human immune system. A normal person can contain a count of 500 to 1500 per cubic millimetre. A person reaches the last stage of HIV—that is, AIDs—when the count reduces to less than 200 per cubic millimetre. In the removal stage, full recovery from HIV/AIDS is not possible; one may just reduce the viral load in the immune system so that it will not be allowed to reach the AIDS stage [15–18].

The following notations are employed in this study:

Let $x_1^{(r)}$, $x_2^{(r)}$ and x_R represent the number of individuals in the susceptible, infectible and removal stages in the transmission of HIV. Here we consider the transmission of HIV between the patches of susceptible and infectible stages. β_{rj} is the transmission rate between infected male to susceptible female—that is, infectious rate and γ are the proportional constants called as removal rate. $n_M^{(j)}$ is the sum of susceptible and infectible males. In the system of Equations (6.1), the susceptible $x_1^{(r)}$ and infectious $x_2^{(r)}$ do not depend on the removal of x_R.

The spread of HIV is presumed to be governed by means of the following rules:

- The populace stays at a fixed level N in the time interval under consideration. That is to say, disregard births and deaths from causes irrelevant to the illnesses under consideration.
- The rate of progress occurring in susceptible populace is proportional to the product of the contributors $x_1^{(r)}$ and people who are removed from suspected $x_1^{(r)}$ at the rate proportional to the size of $x_1^{(r)}$.
- Removal rate of individuals from the infectious at the rate is proportional to the size of $x_2^{(r)}$.

From the above rules, it follows that $x_1^{(r)}$, $x_2^{(r)}$ and x_R satisfy the following system of differential equations [19]:

$$\frac{dx_1^{(r)}}{dt} = -\sum_{j=1}^{k} \beta_{rj} \frac{y_2^{(j)}}{n_M^{(j)}} x_1^{(r)} x_2^{(r)}$$

$$\frac{dx_2^{(r)}}{dt} = \sum_{j=1}^{k} \beta_{rj} \frac{y_2^{(j)}}{n_M^{(j)}} x_1^{(r)} x_2^{(r)} - \gamma x_2^{(r)} \tag{6.1}$$

$$\frac{dx_R}{dt} = \gamma x_2^{(r)}$$

Let us focus attention on the particular subsystem of equations:

$$\frac{dx_1^{(r)}}{dt} = -\sum_{j=1}^{k} \beta_{rj} \frac{y_2^{(j)}}{n_M^{(j)}} x_1^{(r)} x_2^{(r)}$$

$$\frac{dx_2^{(r)}}{dt} = \sum_{j=1}^{k} \beta_{rj} \frac{y_2^{(j)}}{n_M^{(j)}} x_1^{(r)} x_2^{(r)} - \gamma x_2^{(r)} \tag{6.2}$$

Observe that $x_1^{(r)} + x_2^{(r)} + x_R = N$

From Equation (6.2), one gets

$$\frac{dx_2^{(r)}}{dx_1^{(r)}} = \frac{\left(\sum_{j=1}^{k} \beta_{rj} \dfrac{y_2^{(j)}}{n_M^{(j)}} x_1^{(r)} x_2^{(r)} - \gamma x_2^{(r)} \right)}{\left(-\sum_{j=1}^{k} \beta_{rj} \dfrac{y_2^{(j)}}{n_M^{(j)}} x_1^{(r)} x_2^{(r)} \right)}$$

$$= -1 + \left[\frac{\gamma}{\sum_{j=1}^{k} \beta_{rj} \dfrac{y_2^{(j)}}{n_M^{(j)}} x_1^{(r)}} \right] \tag{6.3}$$

$$\frac{dx_2^{(r)}}{dx_1^{(r)}} = -1 + \rho \left(\frac{1}{x_1^{(r)}} \right)$$

where

$$\rho = \frac{\gamma}{\sum_{j=1}^{k} \beta_{rj} \dfrac{y_2^{(j)}}{n_M^{(j)}} x_1^{(r)}}$$

Solving the differential Equation (6.3), we get

$$x_2^{(r)} x_1^{(r)} = x_2^{(r)}(0) + x_1^{(r)}(0) - x_1^{(r)} + \rho log \left(\frac{x_2^{(r)}}{x_2^{(r)}(0)} \right) \tag{6.4}$$

Computation of Threshold Rate for the Spread of HIV

where $x_1^{(r)}(0)$ and $x_2^{(r)}(0)$ are the number of susceptible and infective individuals at the initial time $t = t_0$ and $\rho = \dfrac{\gamma}{\sum_{j=1}^{k} \beta_{rj} \dfrac{y_2^{(j)}}{n_M^{(j)}} x_1^{(r)}}$.

6.6 ANALYSIS OF THE BEHAVIOUR OF THE HIV–SIR MODEL

To analyse the behaviour of the system represented by (6.4), compute

$$(x_1^{(r)})'(x_2^{(r)}) = -1 + \frac{\rho}{x_2^{(r)}}$$

The analysis pertaining to the above computation is as follows:

- If the value of parameter $x_2^{(r)} > \rho$, then the quantity $-1 + \frac{\rho}{x_2^{(r)}}$ is negative.
- The quantity $-1 + \frac{\rho}{x_2^{(r)}}$ becomes positive when the value of parameter $x_2^{(r)} < \rho$.
- If the value of the parameter $x_2^{(r)} > \rho$, then the parameter $x_2^{(r)}$ is a decreasing function.
- Finally, observe that $x_2^{(r)}(0) = -\infty$ and $x_2^{(r)}(x_1^{(r)}) = x_2^{(r)}(0) > 0$.

Hence there exists a unique point $x_1^{(r)}(\infty)$ with $0 < x_1^{(r)}(\infty) < x_1^{(r)}(0)$ such that $x_2^{(r)}(x_1^{(r)}(\infty)) = 0$ and $x_2^{(r)}(x_1^{(r)}) > 0$ for $x_1^{(r)}(\infty) < x_2^{(r)} < x_2^{(r)}(0)$.

It is seen that

$$\frac{dx_1^{(r)}}{dt} \text{ and } \frac{dx_2^{(r)}}{dt} \text{ vanish when } x_2^{(r)} = 0.$$

Therefore it follows that the point $(x_1^{(r)}(\infty), 0)$ is an equilibrium point of the system represented by Equation (6.2).

6.7 THE SPREAD OF HIV WITHIN THE POPULATION

As t keeps running from t_0 to ∞, the point $(x_1^{(r)}(t), x_2^{(r)}(t))$ goes along the curve (6.4) and it moves alongside the curve towards diminishing of $x_1^{(r)}$, since $x_2^{(r)}(t)$ diminishes monotonically. Hence, if $x_1^{(r)}(0)$ is less than ρ, then $x_2^{(r)}(t)$ and $x_1^{(r)}(t)$ diminish monotonically to zero and $x_2^{(r)}(\infty)$, respectively. Thus, a small patch of infective $x_2^{(r)}(0)$ is inserted into a susceptible $x_1^{(r)}(0)$ with $x_1^{(r)}(0) < \rho$. We have to consider three cases.

Case 1: $x_1^{(r)}(0) < \rho$.
In this case, the HIV will die out very quickly.
Case 2: $x_1^{(r)}(0) > \rho$.
In this case, $x_2^{(r)}(t)$ increments as $x_1^{(r)}(t)$ diminishes to ρ.
Case 3: $x_1^{(r)}(0) = \rho$.
In this case, it reaches a maximum value.

From the above consideration, the following conclusions are drawn.

1. An epidemic will arise if the amount of susceptible populace exceeds the threshold value

$$\rho = \frac{\gamma}{\displaystyle\sum_{j=1}^{k} \beta_{rj} \frac{y_2^{(j)}}{n_M^{(j)}} x_1^{(r)}}.$$

2. The transmission of HIV does not stop even with lack of susceptible populace. It stops only when there is lack of infectious populace. Specifically, some individuals escape from the HIV altogether.

The result (6.1) corresponds to the general observations that an epidemic tends to develop an HIV quickly when the thickness of susceptible population is high and the removable rate is low with solitude and inadequate hospital treatment.

In the event of the number of susceptible individuals $x_1^{(r)}(0)$ exceeds ρ but at the same time very nearer to the threshold value ρ, one may undertake computation of the quantity of individuals who eventually contract the HIV. Specifically, if $x_1^{(r)}(0) - \rho$ is very small compared to ρ, then the number of individuals who eventually contract the HIV is approximately $2(x_1^{(r)}(0) - \rho)$. It is known as the threshold theorem for HIV epidemiology.

In order to attain the threshold value for the HIVSIR model, we establish the following result.

Theorem: Let $x_1^{(r)}(0) = \rho + \gamma$ and expect that the ratio of γ to ρ is extremely small contrasted to 1. Assume that the initial number of infectious individuals $x_2^{(r)}(0)$ is too small. Then the number of individuals who eventually contact HIV is 2γ approximately. As ca consequence, the level of susceptible individuals is diminished to a point as far below the threshold as it was originally.

Proof: From the system of equations described in Section 6.5, we get

$$x_2^{(r)} x_1^{(r)} = x_2^{(r)}(0) + x_1^{(r)}(0) - x_1^{(r)} + \rho log\left(\frac{x_2^{(r)}}{x_2^{(r)}(0)}\right)$$

Let 't' approach infinity. Then the above equation gives

$$0 = x_2^{(r)}(0) + x_1^{(r)}(0) - x_1^{(r)}(\infty) + \rho ln\frac{x_2^{(r)}(\infty)}{x_2^{(r)}(0)}$$

When $x_2^{(r)}(0)$ is very small compared to $x_1^{(r)}(0)$, one can omit the parameter $x_2^{(r)}(0)$ from the above equation. Thus we get the relation

Computation of Threshold Rate for the Spread of HIV

$$
\begin{aligned}
0 &= x_1^{(r)}(0) - x_1^{(r)}(\infty) + \rho \ln \frac{x_2^{(r)}(\infty)}{x_2^{(r)}(0)} \\
&= x_1^{(r)}(0) - x_1^{(r)}(\infty) + \rho \ln \left[\frac{x_1^{(r)}(0) - (x_1^{(r)}(0) - x_1^{(r)}(\infty))}{x_1^{(r)}(0)} \right] \\
&= x_1^{(r)}(0) - x_1^{(r)}(\infty) + \rho \ln \left[1 - \left(\frac{x_2^{(r)}(0) - x_2^{(r)}(\infty)}{x_2^{(r)}(0)} \right) \right]
\end{aligned}
$$

Now, if $x_1^{(r)}(0) - \rho < \rho$ then the difference $x_1^{(r)}(0) - x_1^{(r)}(\infty)$ will be very small compared to $x_1^{(r)}(0)$.

Hence by Taylor series expansion, it is truncated as

$$
\ln \left[1 - \left(\frac{x_1^{(r)}(0) - x_1^{(r)}(\infty)}{x_1^{(r)}(0)} \right) \right] = - \left(\frac{x_1^{(r)}(0) - x_1^{(r)}(\infty)}{x_1^{(r)}(0)} \right) - \frac{1}{2} \left(\frac{x_1^{(r)}(0) - x_1^{(r)}(\infty)}{x_1^{(r)}(0)} \right)^2 + \cdots
$$

After two terms, we have

$$
(x_1^{(r)}(0) - x_1^{(r)}(\infty)) - \rho \left(\frac{x_1^{(r)}(0) - x_1^{(r)}(\infty)}{x_1^{(r)}(0)} \right) - \frac{\rho}{2} \left(\frac{x_1^{(r)}(0) - x_1^{(r)}(\infty)}{x_1^{(r)}(0)} \right)^2 = 0
$$

$$
(x_1^{(r)}(0) - x_{1(\infty)}^{(r)}) \left[1 - \frac{\rho}{x_1^{(r)}(0)} - \frac{\rho}{2x_1^{(r)}(0)^2} (x_1^{(r)}(0) - x_1^{(r)}(\infty)) \right] = 0
$$

By simplifying the above equation, one obtains

$$
\left[1 - \frac{\rho}{x_1^{(r)}(0)} - \frac{\rho}{2x_1^{(r)}(0)^2} (x_1^{(r)}(0) - x_1^{(r)}(\infty)) \right] = 0
$$

Therefore,

$$
\begin{aligned}
x_1^{(r)}(0) - x_1^{(r)}(\infty) &= \frac{2x_1^{(r)}(0)^2}{\rho} \left(1 - \frac{\rho}{x_1^{(r)}(0)} \right) \\
&= \frac{2(\rho + \gamma)}{\rho} \left(1 - \frac{\rho}{\rho + \gamma} \right) \\
&= 2 \frac{\gamma}{\rho} \\
&\cong 2\gamma
\end{aligned}
$$

Hence the number of individuals who ultimately contact HIV is approximately 2γ.

6.8 THE RATE OF REMOVAL OF HIV INFECTION

During the span of HIV epidemic, it is difficult to precisely find out the number of recent infectives, since the main infectives who can be perceived and eliminated from circulation are those who seek useful clinical recourse. In order to analyse the outcome anticipated by HIV–SIR model with the actual epidemic, $\frac{dx_R}{dt}$ is evaluated as a function of time. We have

$$\frac{dx_R}{dt} = \gamma x_2^{(r)} = \gamma(N - x_R - x_1^{(r)})$$

This implies that

$$\frac{dx_1^{(r)}}{dx_R} = \frac{-\sum\limits_{j=1}^{k} \beta_{rj} \dfrac{y_2^{(j)}}{n_M^{(j)}} x_1^{(r)}}{\gamma}$$

$$= -\frac{x_1^{(r)}}{\rho}$$

$$dx_1^{(r)} = -\frac{1}{\rho} x_1^{(r)} dx_R$$

$$x_1^{(r)}(x_R) = x_1^{(r)}(0) e^{\frac{-x_R}{\rho}}$$

and

$$\frac{dx_R}{dt} = \gamma \left[N - x_R - x_1^{(r)}(0) e^{\frac{-x_R}{\rho}} \right] \tag{6.5}$$

The above equation can be solved by the method of variables separable, but it is not possible to obtain an explicit solution. It is noted that the quantity $\frac{x_R}{\rho}$ is small when the HIV epidemic is not too large. One may therefore consider trimming the Taylor series for the expansion of the term $e^{\frac{-x_R}{\rho}}$. Applying the above series with the approximation of the terms with higher powers in Equation (6.5), one gets

$$e^{\frac{-x_R}{\rho}} = 1 - \left(\frac{x_R}{\rho} \right) + \frac{1}{2} \left(\frac{x_R}{\rho} \right)^2 + \cdots$$

Computation of Threshold Rate for the Spread of HIV 107

After three terms, one is led to the approximation

$$
\begin{aligned}
\frac{dx_R}{dt} &= \gamma \left[N - x_R - x_1^{(r)}(0) \left[1 - \left(\frac{x_R}{\rho} \right) + \frac{1}{2} \left(\frac{x_R}{\rho} \right)^2 \right] \right] \\
&= \gamma \left[N - x_R - x_1^{(r)}(0) + \frac{x_R x_1^{(r)}(0)}{\rho} - \frac{1}{2} \left(\frac{x_R}{\rho} \right)^2 x_1^{(r)}(0) \right] \\
&= \gamma \left[N - x_1^{(r)}(0) + \left(\frac{x_1^{(r)}(0)}{\rho} - 1 \right) x_R - \frac{x_1^{(r)}(0)}{2} \left(\frac{x_R}{\rho} \right)^2 \right]
\end{aligned}
\tag{6.6}
$$

The solution of the equation is obtained as

$$
x_R(t) = \frac{\rho^2}{x_1^{(r)}(0)} \left[\frac{x_1^{(r)}(0)}{\rho} - 1 + \alpha \, \tanh \left(\frac{1}{2} \alpha \gamma t - \phi \right) \right]
\tag{6.7}
$$

where

$$
\alpha = \left[\left(\frac{x_1^{(r)}(0)}{\rho} - 1 \right)^2 + \frac{2 x_1^{(r)}(0) \left(N - x_1^{(r)}(0) \right)}{\rho^2} \right]^{\frac{1}{2}} \quad and
$$

$$
\phi = \tanh \frac{1}{\alpha} \left(\frac{x_1^{(r)}(0)}{\rho} - 1 \right).
$$

6.9 ANALYSIS OF LOCAL ASYMPTOTIC STABILITY

In this section, the following theorem has been derived for the analysis of the local asymptotic stability. Consider the system of differential equations for the transmission of HIV in mobile heterosexual population given by the SIR model

$$
\begin{aligned}
\frac{dx_1^{(r)}}{dt} &= - \sum_{j=1}^{k} \beta_{rj} \frac{y_2^{(j)}}{n_M^{(j)}} x_1^{(r)} x_2^{(r)} \\
\frac{dx_2^{(r)}}{dt} &= \sum_{j=1}^{k} \beta_{rj} \frac{y_2^{(j)}}{n_M^{(j)}} x_1^{(r)} x_2^{(r)} - \gamma x_2^{(r)} \\
\frac{dx_R}{dt} &= \gamma x_2^{(r)}
\end{aligned}
\tag{6.8}
$$

For this model, we have the following result:

Theorem: The system of differential equations provided by Equation (6.8) is locally asymptotically stable in the positive octant.

Proof: To examine the stability of the system, let us take up the relation

$$\nabla = \frac{\partial}{\partial x_1^{(r)}}(p_1\theta) + \frac{\partial}{\partial x_2^{(r)}}(p_2\theta) + \frac{\partial}{\partial x_R}(p_3\theta)$$

In the background of the theorem on the criterion of divergence, let us consider the relation

$$\theta\left(x_1^{(r)}, x_2^{(r)}, x_R\right) = \frac{1}{x_1^{(r)2} x_2^{(r)2} x_R}$$

where $\theta\left(x_i^{(r)}, x_R\right) > 0$ if $x_i > 0$ and $x_R > 0$

The SIR model concerning the system of differential equations for the transmission of HIV in mobile heterosexual population is provided by

$$p_1 = -\sum_{j=1}^{k} \beta_{rj} \frac{y_2^{(j)}}{n_M^{(j)}} x_1^{(r)} x_2^{(r)}$$

$$p_2 = \sum_{j=1}^{k} \beta_{rj} \frac{y_2^{(j)}}{n_M^{(j)}} x_1^{(r)} x_2^{(r)} - \gamma x_2^{(r)}$$

$$p_3 = \gamma x_2^{(r)}$$

For evaluating the value of ∇ along the trajectories, we have

$$p_1\theta = \frac{-\sum_{j=1}^{k} \beta_{rj} \frac{y_2^{(j)}}{n_M^{(j)}} x_1^{(r)} x_2^{(r)}}{x_1^{(r)2} x_2^{(r)2} x_R}$$

$$p_2\theta = \frac{\sum_{j=1}^{k} \frac{\left(\beta_{rj} y_2^{(j)} x_1^{(r)} - \gamma\right) x_2^{(r)}}{n_M^{(j)}}}{x_1^{(r)2} x_2^{(r)2} x_R}$$

$$p_3\theta = \frac{\gamma x_2^{(r)}}{x_1^{(r)2} x_2^{(r)2} x_R} = \frac{\gamma}{x_1^{(r)2} x_2^{(r)} x_R}$$

Computation of Threshold Rate for the Spread of HIV 109

$$\frac{\partial(p_1\theta)}{\partial x_1^{(r)}} = \frac{\displaystyle\sum_{j=1}^{k}\beta_{rj}\frac{y_2^{(j)}}{n_M^{(j)}}}{x_1^{(r)2}x_2^{(r)}x_R}$$

$$\frac{\partial(p_2\theta)}{\partial x_2^{(r)}} = \frac{-\displaystyle\sum_{j=1}^{k}\left(\beta_{rj}\frac{y_2^{(j)}}{n_M^{(j)}}x_1^{(r)}-\gamma\right)}{x_1^{(r)2}x_2^{(r)2}x_R}$$

$$\frac{\partial(p_3\theta)}{\partial x_R} = -\frac{\gamma}{x_1^{(r)2}x_2^{(r)}x_R^2}$$

With the above relations, we get

$$\nabla = -\left(\frac{\gamma}{x_1^{(r)2}x_2^{(r)}x_R} + \frac{\displaystyle\sum_{j=1}^{k}\beta_{rj}\frac{y_2^{(j)}}{n_M^{(j)}}x_1^{(r)}-\gamma}{x_1^{(r)2}x_2^{(r)2}x_R} - \sum\beta_{rj}\frac{y_2^{(j)}}{x_1^{(r)2}x_2^{(r)}x_R}\right)$$

$$< 0$$

Consequently, it follows by Benedixon-Dulac criterion that there is no limit cycle in the first octant. Hence the system of equations is locally asymptotically stable in the positive octant.

This result can be ascertained in another way also. We have

$$\psi\left(x_1^{(r)},x_2^{(r)},x_R\right) = \frac{dp_1}{dx_1^{(r)}} + \frac{dp_2}{dx_2^{(r)}} + \frac{dp_3}{dx_R}$$

$$= -\sum_{j=1}^{k}\beta_{rj}\frac{y_2^{(j)}}{n_M^{(j)}}x_2^{(r)} + \sum_{j=1}^{k}\beta_{rj}\frac{y_2^{(j)}}{n_M^{(j)}} - \gamma + 0$$

$$\psi = -\gamma$$
$$\neq 0$$

Hence by Bendixon criteria theorem, there is no closed trajectories. This implies that the limit cycle does not exist. Thus it is ascertained that the system of equations is locally asymptotically stable.

6.10 NUMERICAL SIMULATION

Using MATLAB® ODE solver, a numerical simulation has been carried out for the model discussed in this chapter.

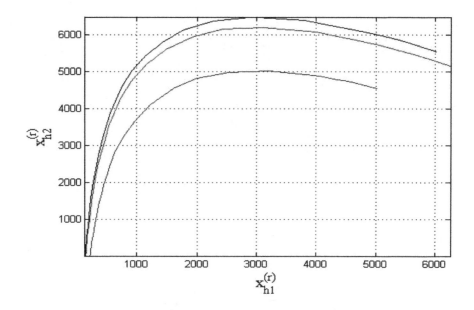

FIGURE 6.2 Solution orbits for initial values.

For simulation, the following sets of values have been taken:

$$I.\ x_1^{(r)} = 5013,\ x_2^{(r)} = 4558\ \text{and}\ x_R = 4218$$
$$II.\ x_1^{(r)} = 6013,\ x_2^{(r)} = 5558\ \text{and}\ x_R = 3218$$
$$III.\ x_1^{(r)} = 6258,\ x_2^{(r)} = 5158\ \text{and}\ x_R = 4218$$

Figure 6.2 represents the solution orbits for the given initial values. The stability curve is provided by Figure 6.3.

From the simulation study, the following results emanate:

1. When the value of t is finite, $x_1^{(r)}$, $x_2^{(r)}$ and x_R monotonically decreases to zero.
2. The spread of the disease may not be stopped by susceptible. But it can be controlled by infectible.

The simulation study confirms our result pertaining to the threshold rate of the system of HIV in a mobile heterosexual population.

6.11 CONCLUSION

In this chapter, the mathematical model for a mobile heterosexual population has been considered. The threshold rate for the transmission of HIV in a mobile heterosexual population has been analysed. HIV transmission of heterosexual female

Computation of Threshold Rate for the Spread of HIV

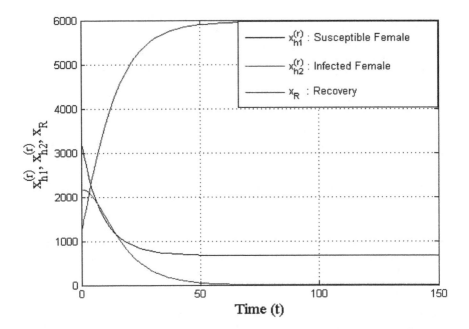

FIGURE 6.3 Stability curve.

population is considered by means of a simple SIR model. The threshold value of the number of individuals who are eventually prevented from passing on to the AIDS stage is obtained approximately as 2γ. An epidemic occurs when the number of susceptible females and males exceeds the threshold value. The removal rate of last-stage HIV infection is derived. The results from this study would benefit the AIDS/HIV prevention agencies to augment their efficiency in the prevention from susceptible stage to infection stage. The findings in this work would motivate future researchers to take up studies on the construction of mathematical models to identify crucial parameters to prevent the spread of HIV/AIDS.

REFERENCES

1. Hethcote, H. W., (2000), 'The Mathematics of Infectious Diseases', Society for Industrial and Applied Mathematics (SIAM-Review), Vol. 42, pp. 599–653.
2. Waziri, A. S., Massawe, E. S., and Makinde, O. D., (2012), 'Mathematical Modelling of HIV/AIDS Dynamics with Treatment and Vertical Transmission', Applied Mathematics, Vol. 2, pp. 77–89.
3. Bashiru, K. A., and Fasoranbaku, O. A., (2009), 'Statistical Modeling of Mother-to-Child and Heterosexual Modes of Transmission of HIV/AIDS Epidemic', Pacific Journal of Science and Technology, Vol. 10, pp. 966–979.
4. Nah, K., Nishiura, H., Tsuchiya, N., Sun, X., Asai, Y., and Imamura, A., (2017), 'Test-and-Treat Approach to HIV/AIDS: A Primer for Mathematical Modeling', Theoretical Biology and Medical Modelling, Vol. 14, pp. 1–11.

5. Ma, J., and Ma, Z., (2006), 'Epidemic Threshold Conditions for Seasonally Forced SEIR Models', Mathematical Biosciences, Vol. 3, pp. 161–172.
6. Heesterbeek, J. A. P., (2002), 'A Brief History of R_0 and a Recipe for Its Calculation', Acta Biotheoretica, Vol. 50, pp. 189–204.
7. Hefferman, J., Smith, R., and Wahl, I., (2005), 'Perspectives on the Basic Reproduction Ratio', Journal of the Royal Society Interface, Vol. 2, pp. 281–293.
8. Chunyan J., Jiang, D., (2014), 'Threshold Behaviour of a Stochastic SIR Model', Applied Mathematical Modelling, Vol. 38, pp. 5067–5079.
9. Franceschetti, P. A., (2008), 'Threshold Behaviour of a SIR Epidemic Model with Age Structure and Immigration', Journal of Mathematical Biology, Vol. 57, pp. 1–27.
10. Wang, L., Teng, Z., Tang, T., and Li, Z., (2017), 'Threshold Dynamics in Stochastic SIRS Epidemic Models with Nonlinear Incidence and Vaccination', Computational and Mathematical Methods in Medicine, Vol. 2017, pp. 1–20.
11. Miao, A., Zhang, J., Zhang, T., and Pradeep, B. G. S. A., (2017), 'Threshold Dynamics of a Stochastic SIR Model with Vertical Transmission and Vaccination', Computational and Mathematical Methods in Medicine, Vol. 2017, pp. 1–10.
12. Rodrigues., H. S., (2016), 'Applications of SIR Epidemiological Model: New Trends', International Journal of Applied Mathematics and Informatics, Vol. 10, pp. 92–97.
13. Mbah, G. C. E., Omale., D., and Adejo., B. O., (2014), 'A SIR Epidemic Model for HIV/ AIDS Infection', International Journal of Scientific and Engineering Research, Vol. 5, pp. 479–484.
14. Miller, J. C., (2017), 'Mathematical Models of SIR Disease Spread with Combined Non-Sexual and Sexual Transmission Routes', Infectious Disease Modelling, Vol. 2, pp. 35–55.
15. Ali, M. A., Rafiq, M., and Ahmad, M. O., (2019), 'Numerical Analysis of a Modified SIR Epidemic Model with The Effect of Time Delay', Journal of Mathematics, Vol. 51, pp. 79–90.
16. Chauhan, S., Misra, O. P., and Dhar, J., (2014), 'Stability Analysis of SIR Model with Vaccination', American Journal of Computational and Applied Mathematics, Vol. 4, pp. 17–23.
17. Jiang, D., Ji, C., Shi, N., and Yu, J., (2010), 'The Long Time Behavior of SIR Epidemic Model with Stochastic Perturbation', Journal of Mathematical Analysis and Applications, Vol. 372, pp. 162–180.
18. Liu, M., Bai, C., and Wang, K., (2014), 'Asymptotic Stability of a Two Group Stochastic SEIR Model with Infinite Delays', Communications in Nonlinear Science and Numerical Simulation, Vol. 19, pp. 3444–3453.
19. Sani, A., Kroese, D. P., and Pollett, P. K., (2007), 'Stochastic Models for the spread of HIV in a mobile heterosexual population', Mathematical Biosciences, Vol. 208, pp. 98–124.

7 Application of Soft Computing Techniques to Heart Sound Classification
A Review of the Decade

Babita Majhi and Aarti Kashyap
Department of CSIT
Guru Ghasidas Vishwavidyalaya
Bilaspur, Chhattisgarh, India

CONTENTS

7.1 Introduction 113
7.2 Related Literature Review 114
7.3 Steps for Heart Sound Classification 123
    7.3.1 Pre-Processing 123
    7.3.2 Feature Extraction 127
    7.3.3 Classification... 127
7.4 Research Gap.. 129
7.5 Conclusion 133
References... 134

7.1 INTRODUCTION

Heart sound classification plays an important role in the diagnosis and prevention of cardiovascular disease and is used for automatic heart sound auscultation and cardiac monitoring [1]. As indicated by the World Health Organization, nearly 17.5 million individuals around the world have died because of cardiovascular illness, which is 31–32% of all deaths and its rate is expanding quickly [2]. The heart is one of the most significant organs of the human body and conveys blood to all parts. The heart works like a siphon and pulsates 72 times each moment for a normal individual under ordinary conditions [3]. The heart pumps blood through a network of arteries and veins called the cardiovascular system. The human heart has four chambers: the right atrium, the right ventricle, the left atrium and the left ventricle. The human heart performs the duties in two cycles: systole and diastole. The contraction of heart is known as systole and the relaxation of

114   Soft Computing Applications and Techniques in Healthcare

heart is known as diastole. The heart sound can be produced using two sounds, 'lub' and 'dub', in sequence that occur due to the closing of the valves of the heart [4]. The abnormal sound is produced due to damaged valves. Because of the disorder of the heart valve, the common disease of the heart occurs. Some of the diseases that occur are myocardial infarction (heart attack), congestive heart failure, heart murmur, coronary artery disease, heart valve disease, stable angina pectoris, unstable angina, pectoris and arrhythmia. The primary method is the auscultation used by the physicians to differentiate between normal and abnormal heart sounds. Any disorders can be detected by the physicians after listening these sounds using the stethoscope, digital applications and so on [4].

Traditionally, cardiologists use stethoscopes for examination of heart sounds. The accuracy of heart sound classification is based on the experience and skill of the physicians. But this manual clinical process is time-consuming and costly. To alleviate these limitations, recently a computer-based automatic computer assist tool is recommended for detection of abnormal heart sound. Hence this is becoming an emerging research for the biological signal processing and machine learning groups as it is computer based. Soft computing is one of the problem-solving approaches used to solve real life complex problems in the field of science and technology. Applications of various soft computing techniques such as artificial neural network, fuzzy logic and evolutionary computing have been extensively used in the medical diagnosis. Various soft computing techniques are also applied by the researchers in the field of classification of heart sound.

The main objective of this chapter is to provide a systematic review of different existing approaches for the classification of heart sounds of the last 10 years from 2008 to 2018. Also, this chapter will provide the details about the databases, techniques applied in designing models, classifiers used, extract features, domain analysis and performance comparison between review papers. Lots of research has already been done on heart sound classification. However, there is still work to be done in this area through the development of different algorithms and techniques. In particular, the development of some smart mobile applications will be helpful in the improvement of cardiovascular disease diagnosis.

The rest of this chapter is organised in the following sections. Section 7.2 provides the systematic related literature review. Section 7.3 describes the three steps of heart sound classification: preprocessing, feature extraction and classification. The research gap (proposed work and future work) is discussed in Section 7.4. Finally, Section 7.5 is a conclusion of this chapter.

7.2 RELATED LITERATURE REVIEW

Reviewing literature is a key part of research, as it works as a guidepost, not only because it shows the quantum of work done in the field but also because it enables us to perceive the gap and lacuna in the related field of research. It helps in understanding the potentiality of the problem at hand and ensures the evidence of unnecessary duplication. The purposes of the survey of related literature are to locate comparative data useful in the interpretation of results and to provide ideas, theories and explanations in solving the problem.

A systematic review of papers from 2008 to 2018 describing various works done in the field of heart sound (HS) classification are shown in Table 7.1.

Application of Soft Computing Techniques to Heart Sound Classification   115

TABLE 7.1

Review of Heart Sound Classification Techniques and Their Performance Comparison

Reference	Year	No. of Signals	Domain Analysis	Classifiers	Results Based on Reported Metrics in Percentage
Bin Xiao et al. [5]	2018	Total 3153 HS, from where 2488 normal and 665 abnormal heart sound taken from Physionet/CinC2016	2-D time-frequency domain	Neural network (NN)	Accuracy = 93%, specificity = 95% and sensitivity = 86%
Shahid Ismail Malik et al. [3]	2018	Dataset— Normal HS = 21 auscultations labelled and unlabelled signals = 10	Time domain	History-based classifier	Accuracy = 85%, sensitivity = 97% and precision = 88%
Shanti R. Thiyagaraja et al. [6]	2018	Dataset consists of normal HS = 21 and abnormal HS = 41 (total training data = 223 and testing data = 199)	Time and frequency domain	Hidden Markov model (HMM)	Accuracy of normal HS 76.19% and normal and abnormal HS 80.76% and 92.68%
Fatima Chakir et al. [7]	2018	Dataset A and B taken from PASCAL HS challenge. Training set-124 signals and 312 signals and test set = 52 and 195 recording for dataset A and B.	Time domain	K-nearest neighbour (KNN) and discriminant analysis (DA)	Total precision for dataset A and B is 2.96 and 1.58.
B. Bozkurt et al. [8]	2018	Data taken from UoC-murmur database and PhysioNet Challenge 2016 database	Time-frequency domain	Convolutional neural network (CNN)	Sensitivity = 84.5%, specificity = 0.785 and accuracy = 0.815

(Continued)

TABLE 7.1 *(Continued)*
Review of Heart Sound Classification Techniques and Their Performance Comparison

Reference	Year	No. of Signals	Domain Analysis	Classifiers	Results Based on Reported Metrics in Percentage
Qurat-ul-Ain Mubarak et al. [9]	2018	Dataset from Pascal Classifying Heart Sound Challenge	Time and frequency domain	Support vector machine (SVM)	Accuracy = 91%
Hong Tang et al. [10]	2018	Database from PhysioNet/ CinC challenge 2016 total of 3153 HS recordings in WAV format from 764 subjects	Time, frequency, energy, spectrum, ceptrum domain	Support vector machine (SVM)	Sensitivity = 88%, specificity = 87% and overall = 88%
Maryam Hamidi et al. [11]	2018	Dataset A, B, C, D, E includes 115, 386, 7, 27, 1088 normal and 287, 103, 22, 28, 103 pathological heart sounds (PCG) signals.	Frequency domain	K-nearest neighbour classifier (KNN) with Euclidian distance	For three datasets overall accuracy 92%, 81% and 98% is achieved.
Kimitake Ohkawa et al. [12]	2018	Samples for unhealthy patients: (L1 = 89(33,37%), L2 = 47(33,70%), L3 = 38(18,47%)) and healthy patients: (L1 = 89(37,42%), L2 = 47(36,77%), L3 = 38(22,58%))	—	Hidden Markov, model (HMM)	Highest classification performance is 89.9%. Healthy and unhealthy patients is 86.6%.
H. M. Fahad et al. [13]	2018	Dataset—150 Normal HS = 150 and abnormal HS = 80	Frequency domain	Inference system using adaptive neuro fuzzy and hidden Markov model (HMM)	Accuracy = 98.7%

(Continued)

Application of Soft Computing Techniques to Heart Sound Classification   117

TABLE 7.1 *(Continued)*

Reference	Year	No. of Signals	Domain Analysis	Classifiers	Results Based on Reported Metrics in Percentage
Wenjie Zhang et al. [14]	2017	Dataset A includes 176 records, dataset B includes 507 records and dataset C includes 3240 records.	Time-frequency domain	Support vector machine (SVM)	Total highest precision for dataset A = 3.17, dataset B = 2.03 by SVM-DM and for dataset C normal precision = 0.96 by DRGE, abnormal precision 0.88 by SS-TD, overall score 0.90 bt SS-TD
Huseyin Coskun et al. [15]	2017	Heart sound data carried out from PASCAL HS database. 45 extra systole HS, 30 of them are training data and remaining 15 is test data.	Time-frequency domain	Artificial neural network (ANN)	Accuracy = 90%
Sunjing et al. [16]	2017	PCG data has 800MB to collect from 1000 records. Normal PCG data = 279MB and abnormal data = 521 MB	Frequency domain	Convolutional neural network (CNN)	CNN achieves 80.2%.
Diogo Marcelo Nogueira1 et al. [17]	2017	Dataset A contains a total of 400 HS recordings and 400 ECG signals.	Time and frequency domain	Support vector machine (SVM), CNN and random forest	Sensitivity = 91.8%, specificity = 82.05, overall 86.96% and accuracy = 86.97%
Mohammad Nassralla et al. [18]	2017	Dataset obtained from PhysioNet Challenge 2016. Total HS recordings is 3126.	Time-frequency domain	Neural network (NN) – random forests, decision trees	Specificity = 98%, sensitivity = 8% and accuracy = 92%

(Continued)

TABLE 7.1 *(Continued)*
Review of Heart Sound Classification Techniques and Their Performance Comparison

Reference	Year	No. of Signals	Domain Analysis	Classifiers	Results Based on Reported Metrics in Percentage
Wenjie Zhang et al. [19]	2017	Dataset A consists of 176 records with 44,100 Hz and dataset B consists of 507 records with 4000 Hz sampling frequency	Time-frequency Domain	Support vector machine (SVM)	Dataset-A: Normal category = 67%, murmur = 91%, extra heart sound = 44% and artifact = 94% by SS-method Dataset-B: Murmur category in SS-method = 66%&SS-PLSR = 65%
Bradley M. Whitaker et al. [20]	2017	Dataset taken from PhysioNet challenge 2016 which includes total 3,153 audio PCG recordings	Time domain	Support vector machine (SVM)	Dataset E achieves score between 93.4% and 96.7%. (MAcc) on a different dataset is 56.4%.
Lubaib P. et al. [21]	2016	The data is obtained from Michigan University website.	Time domain and frequency domain	SVM classifier, KNN, Bayesian and Gaussian mixture model	Highest accuracy 99.98% and 99.99% obtained by SVM and 99.99% accuracy obtained by Bayesian classifier.
Shi-Wen Deng et al. [22]	2016	Dataset A— total 176 records in WAV format (44,100 Hz) and Dataset B—total 656 records in WAV format (4000 Hz) sampling frequency	Frequency domain	Support vector machine (SVM)	Best precisions are 91% and 100%achieved by SVM-DM, Murmur and Extra Sound categories is 92% achieved by SVM-DM with sensitivity 59%. SVM-DM obtained highest F-score 0.74 for dataset A.

(Continued)

Application of Soft Computing Techniques to Heart Sound Classification   119

TABLE 7.1 *(Continued)*

Reference	Year	No. of Signals	Domain Analysis	Classifiers	Results Based on Reported Metrics in Percentage
Heechang Ryu et al. [23]	2016	Training data set (3126 recordings) and its performance were evaluated using the validation dataset (300 recordings) and hidden test dataset	Frequency domain	Convolutional neural network (CNN)	Overall score accuracy 79.5%, sensitivity = 70.8% and specificity = 88.2%
Edmund Kay et al. [24]	2016	Dataset (a, b, c, d, e, f) have validation data (NG = 149, NP = 5, AG = 126, AP = 32) and training data (NG = 486, NP = 39, AG = 468, AP = 63)	Time-frequency Domain	Drop connect classifier	Classification score = 84.1% with a variance of 2.9%. Test data is 85.2%
Tanachat Nilanon et al. [25]	2016	Training samples from around 3000 to 60,000	Time-frequency Domain	Baseline classifier includes logistic regression, SVM and random forests and CNN	A score of 81.3% with sensitivity = 73.5% and specificity = 89.2%
Gari D. Clifford et al. [26]	2016	Database includes 4430 recordings collected from both healthy subjects and patients, 1072 patients, total 233,512 HS.	Time domain	Benchmark classifier	Sensitivity = 94.24%, specificity = 95.21% and MAcc = 86.02%
Christian Thomae et al. [27]	2016	PhysioNet Challenge 2016.Total training data: 3153 HS recordings	Time-frequency domain	ANN–multi-layer perceptron	A training/validation split of 85%/15%, result obtained is 0.89. Non-revised score = 77% and revised score = 82.7%.

(Continued)

TABLE 7.1 *(Continued)*
Review of Heart Sound Classification Techniques and Their Performance Comparison

Reference	Year	No. of Signals	Domain Analysis	Classifiers	Results Based on Reported Metrics in Percentage
Bradley M. Whitaker et al. [28]	2016	The 3153 labelled audio recordings data is taken from PhysioNet/ CinC Challenge 2016.	Sparse domain	Support vector machine (SVM)	Achieves a cross-validation score of 86.52%; sensitivity = 86.69% and Specificity = 86.34%
Soo-Kng Teo et al. [29]	2016	Dataset taken from PhysioNet/ Computing in Cardiology. Challenge 2016	Frequency domain	Neural network (NN)	Accuracy of sensitivity = 74.7%, specificity = 78.8% and overall = 76.7%
Michael Tschannen et al. [30]	2016	Dataset taken from PhysioNet/ CinC Challenge 2016	Time-frequency domain	Support vector machine (SVM)	Score = 81.2%, sensitivity = 84.8% and specificity = 77.6%
Mohamed Moustafa Azmy [31]	2015	Total no. of features is 40. 90 heart sound for training set and 64 heart sound for test set.	Time-frequency domain	Support vector machine (SVM)	The obtained results of accuracy is 92.29%, specificity is 95.38% and sensitivity is 90%.
Simarjot Kaur Randhawa et al. [32]	2015	Total samples = 144, normal = 60, diastolic murmur = 45 and systolic murmur = 39 signals	Time domain, frequency domain and statistical domain	KNN, fuzzy KNN and ANN	KNN and fuzzy KNN classifiers both have the highest accuracy of 99.6%.
V. Nivitha Varghees et al. [33]	2015	SNR heart sounds (20,15,10,5) and heart murmur(HM) sub-bands	Frequency domain	Decision rule classifier	$P_{ms} = 100\%$, for HS $P_{hs} = 97.33\%$ and falsely detecting $P_{fs} = 1.33\%$ Accuracy from 82.76% to 100%

(Continued)

TABLE 7.1 *(Continued)*

Reference	Year	No. of Signals	Domain Analysis	Classifiers	Results Based on Reported Metrics in Percentage
Fatemah Safara [34]	2015	Total = 59 HS, normal HS = 16 and pathological HS = 43 (19 MR, 14 AS, 10 AR)	Frequency domain	Support vector machine (SVM)	99.39% accuracy is achieved by CT_LDB, 98.78% by CT_MBS and 96.95% by CT_BBS.
Shivnarayan Patidar et al. [35]	2014	Dataset consists of 50 abnormal cardiac sound signals.	Time domain	Least square- support vector machine (LS-SVM)	Accuracy = 94.01%
Laurentius Kuncoro ProboSaputra et al. [36]	2014	Dataset is taken from Michigan HS database. Abnormal HS = 13 types from 13 categories	Frequency domain	Artificial neural network (ANN)	Classifier system achieves the accuracy of 92.31%.
Grzegorz Redlarski et al. [37]	2014	Total 12 different sets of waveforms (normal and abnormal HS)	Time and frequency domain	SVM-MCS (modified cuckoo search)	The developed system achieved accuracy of above 93%.
Fatemeh Safara et al. [38]	2013	Total = 59 HS including normal = 16 and pathological heart sounds = 43 (19 MR, 14 AS and 10 AR)	Frequency domain	Support vector machine (SVM)	Average accuracy = 97.56%
Elsa Ferreira Gomes et al. [39]	2013	PASCAL challenge Total 312 auscultations taken for dataset B, three classes: normal = 200, murmur = 66 cases and extra systole = 46	Time Domain	Decision tree and multi-layer perceptron	The highest accuracy 72.76% is obtained by random forest algorithm.

(Continued)

TABLE 7.1 *(Continued)*
Review of Heart Sound Classification Techniques and Their Performance Comparison

Reference	Year	No. of Signals	Domain Analysis	Classifiers	Results Based on Reported Metrics in Percentage
M. Markaki et al. [40]	2013	The database consists of 25 cases with innocent murmur, normal and 25 patients with abnormal systolic murmurs.	Time-frequency domain	Support vector machine (SVM)	The proposed system achieves score of 95.8% and 96.07%.
Elsa Ferreira Gomes et al. [41]	2012	Dataset A and B taken from Classifying Heart Sound PASCAL Challenge	Frequency domain	Multi-layer perceptron (MLP)	The total highest precision is achieved by MLP for dataset A and B is 2.12 and 1.67.
C. Kwak et al. [42]	2012	Two sets of data, normal category = 80 and abnormal category = 80	Time domain	Support vector machine (SVM)	MLP = 83.1%, RBF = 84.4% & SVM = 85.6%
Fatemeh Safara et al. [43]	2012	Dataset consists of total 350 HS, normal = 50 and murmurs = 300	Time and frequency domain	Decision tree, KNN, Bayes net, MLP and SVM	96.94% accuracy obtained by Bayes Net and 95.33% by SVM
Di Zhang et al. [44]	2011	The database consists of normal = 100 and pathological sounds = 50.	Frequency domain	Linear discriminant analysis (LDA)	97.3% (146/150) and 91.3% (137/150) for HHT + LDA and LDB + LDA
Gur Emre Guraksin et al. [45]	2011	The heart sound taken from Littmann 4100 model electronic stethoscope in WAV format	Time and frequency domain	Least square-support vector machine (LS-SVM)	Accuracy obtained is 96.66%.
Samjin Choi et al. [46]	2010	The database consists of 489 cardiac sound signals, normal = 196 and abnormal = 293.	Frequency domain	Multi-support vector machine (SVM)	Specificity = 99.9%; sensitivity = 99.5%
Guy Amit et al. [47]	2009	Two heart sound datasets—HSPRS (12 subjects) and HSDSE (11 subjects)—were taken as input.	Time-frequency domain	K-nearest neighbour (KNN)	First and second dataset achieved an average accuracy of 82±7% and 86±7%

(Continued)

TABLE 7.1 (Continued)

Reference	Year	No. of Signals	Domain Analysis	Classifiers	Results Based on Reported Metrics in Percentage
Zumray Dokur et al. [48]	2008	140 HS cycles of 14 different heart sounds for classification performance of ISOM and Kohonen's network were taken as input.	Time domain	Kohonen's self- organising map (SOM); network and incremental self- organising map (ISOM)	95% accuracy obtained by ISOM and 75% by Kohonen's network

7.3 STEPS FOR HEART SOUND CLASSIFICATION

Figure 7.1 shows the flowchart of heart sound signal processing which includes preprocessing, extraction of features and classification. Different techniques/methods/approaches are used to improve the accuracy of heart sound signals for preprocessing, extraction of features and classification. Firstly, the signal is pre-processed, then feature extraction occurs and finally the classifier is used for classification.

7.3.1 Pre-Processing

The process of removal of noises is known as preprocessing. Heart sound signals are de-noised, normalised and then classified. Various preprocessing techniques applied to process heart sound signals by various authors are described in Table 7.2.

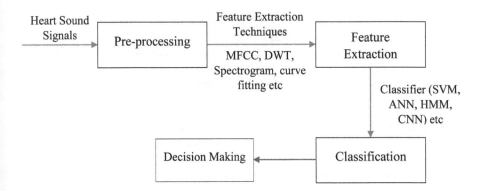

FIGURE 7.1 Flowchart of heart sound classification.

124 Soft Computing Applications and Techniques in Healthcare

TABLE 7.2
Different Pre-Processing and Feature Extraction Techniques Reviewed Year-Wise from 2008 to 2018

Reference	Year	Pre-Processing and Features Extraction Techniques
Bin Xiao et al. [5]	2018	Mel-frequency cepstral coefficients (MFCC) and power spectral density (PSD)
Shahid Ismail Malik et al. [3]	2018	Low-pass filtering, fast Fourier transform (FFT)
Shanti R. Thiyagaraja et al. [6]	2018	Discrete wavelet transform (DWT) and continuous wavelet transform (CWT), fast Fourier transform (FFT), short-time Fourier transform (STFT), mel-frequency cepstral coefficients (MFCC)
Fatima Chakir et al. [7]	2018	Down sampling, envelope extraction using Shannon energy, band-pass filtering and normalisation
B. Bozkurt et al. [8]	2018	Spectrogram, mel-spectrogram, mel-frequency cepstral coefficients (MFCC), sub-band envelope
Qurat-ul-Ain Mubarak et al. [9]	2018	For preprocessing: forward-backward filter, Butterworth band-pass filter, spike removal algorithm; for feature extraction: Shannon energy, discrete wavelet transform
Hong Tang et al. [10]	2018	For pre-processing: spike removal algorithm; for feature extraction: Gaussian window and discrete Fourier transform
Maryam Hamidi et al. [11]	2018	Wavelet transform (WT), entropy, power spectral density (PSD), curve fitting, fractal dimension and mel-frequency cepstral coefficients (MFCC)
Kimitake Ohkawa et al. [12]	2018	Heart sound feature extraction methods
H.M. Fahad et al. [13]	2018	For preprocessing: down-sampling, normalisation. For feature extraction: systole and diastole energy, systole and diastole frequency
Wenjie Zhang et al. [14]	2017	For preprocessing: a band-pass, sixth-order Butterworth filter; for feature extraction: scaled spectrogram and tensor decomposition
Huseyin Coskun et al. [15]	2017	For pre-processing: elliptic filter for noise reduction; for feature extraction: mel-frequency cepstral coefficients (MFCC)
Sunjing et al. [16]	2017	For preprocessing: wavelet threshold; for feature extraction: mel-frequency cepstral coefficients (MFCC)

(Continued)

Application of Soft Computing Techniques to Heart Sound Classification 125

TABLE 7.2 *(Continued)*

Reference	Year	Pre-Processing and Features Extraction Techniques
Diogo Marcelo Nogueira1 et al. [17]	2017	For preprocessing: windowing, fast Fourier transform (FFT), mel-filtering, discrete cousin transform; for feature extraction: mel-frequency cepstral coefficients (MFCC)
Mohammad Nassralla et al. [18]	2017	For preprocessing: low pass filter; for feature: mel-frequency cepstral coefficients (MFCC), wavelet entropy, power spectrum
Wenjie Zhang et al. [19]	2017	For pre-processing: a band-pass, sixth-order Butterworth filter; for feature extraction: discrete wavelet transform, scaled spectrogram, partial least squares regression (PLSR), Shannon energy envelope
Bradley M. Whitaker et al. [20]	2017	For preprocessing: fast Fourier transform (FFT); for feature extraction: sparse coding
Lubaib P. et al. [21]	2016	For feature extraction: mel-frequency cepstral coefficients (MFCC), discrete cousin transform (DCT)
Shi-Wen Deng et al. [22]	2016	For preprocessing: band-pass, zero-phase, Butterworth filter order, average Shannon energy envelope (ASEs), discrete wavelet transform (DWT); tor feature extraction: diffusion maps
Heechang Ryu et al. [23]	2016	For pre-processing: windowed-sinc Hamming filter algorithm; for feature extraction: convolution layers, nonlinear activation function, batch normalisation, max pooling layers
Edmund Kay et al. [24]	2016	For feature extraction: wavelet transform (WT), continuous wavelet transform (CWT), mel-frequency cepstral coefficients (MFCC), inner-beat features, principal component analysis (PCM)
Tanachat Nilanon et al. [25]	2016	For feature extraction: spectrograms, PSD, MFCC
Gari D. Clifford et al. [26]	2016	PhysioNet/computing in cardiology challenge method
Christian Thomae et al. [27]	2016	Gaussian noise, max pooling, batch normalisation, bidirectional gated recurrent units, feed forward attention MLP
Bradley M. Whitaker et al. [28]	2016	For preprocessing: N point FFT; for feature extraction: sparse coding

(Continued)

126 Soft Computing Applications and Techniques in Healthcare

TABLE 7.2 *(Continued)*
Different Pre-Processing and Feature Extraction Techniques Reviewed Year-Wise from 2008 to 2018

Reference	Year	Pre-Processing and Features Extraction Techniques
Soo-Kng Teo et al. [29]	2016	For segmentation: Springer algorithm; for feature extraction: power spectrum analysis, FFT
Michael Tschannen et al. [30]	2016	For feature extraction: 2-D Haar wavelet filter, spectrogram, PSD
Mohamed Moustafa Azmy [31]	2015	For preprocessing: multi-level filter ban; for feature extraction: discrete wavelet transform, new mother wavelet transform
Simarjot Kaur Randhawa et al. [32]	2015	For feature extraction: Spectrum analysers, including SpectraPlus SC and Sigview
V. Nivitha Varghees et al. [33]	2015	Stationary wavelet transform (SWT) and Hilbert phase envelope, relative wavelet sub-band energy ration (RWSER)
Fatemah Safara [34]	2015	Wavelet packet tree (WPT), higher-order cumulants (HOC), cumulant-based trapezoidal best basis selection (CT_BBS)
Shivnarayan Patidar et al. [35]	2014	For feature extraction: constrained tunable-Q wavelet transform, short-time Fourier transform (STFT)
Laurentius Kuncoro Probo Saputra et al. [36]	2014	For preprocessing and feature extraction: wavelet decomposition, autoregressive power spectral density (AR-PSD)
Grzegorz Redlarski et al. [37]	2014	For feature extraction - Modified Linear Predictive Coding Algorithm and for feature extraction
Fatemeh Safara et al. [38]	2013	For pre-processing and feature extraction: multi-level basis selection, multi-level basis selection, wavelet packet decomposition tree
Elsa Ferreira Gomes et al. [39]	2013	For preprocessing and feature extraction: filters and normalised average Shannon energy, MrMotif algorithm
M. Markaki et al. [40]	2013	For feature extraction: short-time Fourier transform (STFT)
Elsa Ferreira Gomes et al. [41]	2012	For preprocessing: band-pass filter (fifth-order Chebyshev type I low-pass filter); for feature extraction: Shannon energy, peak detection algorithm
C. Kwak et al. [42]	2012	For feature extraction: mel-frequency cepstral coefficients (MFCC)
Fatemeh Safara et al. [43]	2012	For preprocessing: band-pass finite-duration impulse response (FIR); for feature extraction: wavelet packet entropy (WPT)

(Continued)

Application of Soft Computing Techniques to Heart Sound Classification   **127**

TABLE 7.2 *(Continued)*

Reference	Year	Pre-Processing and Features Extraction Techniques
Di Zhang et al. [44]	2011	For feature extraction: Hilbert Haung transform (HHT)
Gur Emre Guraksin et al. [45]	2011	For feature extraction: wavelet transform (WT), DWT, CWT, Shannon energy
Samjin Choi et al. [46]	2010	For preprocessing and feature extraction: auto regressive power spectral density (ARPSD), analysis, wavelet based preprocessing, NAR-PSD curve
Guy Amit et al. [47]	2009	For preprocessing: digital band-pass filter; for feature extraction: FFT, STFT, S-transform (S-T), Winger-Ville distribution (WVD) and Choi-Williams distribution (CWD)
Zumray Dokur et al. [48]	2008	For preprocessing: adaptive peak detector; for feature extraction: discrete wavelet transform, wavelet transform

7.3.2 FEATURE EXTRACTION

Feature extraction is a process of transforming the data into a suitable form. It is used to reduce the dimensionality of the data. Also, it removes noise and redundant information. Transforming of data is essential to extract features for the recognition of heart sound signals, where the data are represented in a useful form [14]. Features are extracted using different time and frequency domain techniques, which are listed in Table 7.2.

7.3.3 CLASSIFICATION

Classification is a supervised machine learning technique which is a process of grouping objects into classes. The process of classification has two phases: training and testing. To identify any abnormalities in the heart sound, classification is used. In this chapter, list of different classifiers is used for classification of heart sound signals is described in Table 7.3.

It is observed that mostly various types of neural networks and SVM are used for classification of heart sound signals. The accuracy obtained by various classifiers are 93% by NN [5], 91% by SVM [9], 98.7% by HMM [13], 90% by ANN [15], 92% by NN [18], 99.99% by SVM and Bayes net [21], 92.29% by SVM [32], 99.6% KNN and fuzzy KNN [34], 99.39% by SVM (CT_LDB) [35], 94.01% by LS-SVM [36], 32.31% by ANN [37], 93% by SVM-MCS [38], 97.56% by SVM [40], 96.07% by SVM [43], 96.94% and 95.33% by Bayes net and SVM [45], 96.66% by LS-SVM [48] and 95% by ISOM classifiers. The highest accuracy reported in heart sound is 99.99% [21]. The results exhibit that SVM is one of the mostly used classifier and SVM and Bayes net is the best classifier.

TABLE 7.3
Different Classifiers and Percentage of Accuracy

Reference	Classifiers Used	% of Accuracy
Bin Xiao et al. [5]	Neural network (NN)	93%
Shahid Ismail Malik et al. [3]	History-based classifier	85%
Shanti R. Thiyagaraja et al. [6]	Hidden Markov model (HMM)	80.76%
Fatima Chakir et al. [7]	K-nearest neighbour (KNN) and discriminant analysis (DA)	Not mentioned
B. Bozkurt et al. [8]	Convolutional neural network (CNN)	81.5%
Qurat-ul-Ain Mubarak et al. [9]	Support vector machine (SVM)	91%
Hong Tang et al. [10]	Support vector machine (SVM)	88%
Maryam Hamidi et al. [11]	K-nearest neighbour classifier (KNN) with Euclidian distance	For three datasets: 92%, 81% and 98%
Kimitake Ohkawa et al. [12]	Hidden Markov model (HMM)	89.9%
H.M. Fahad et al. [13]	Inference system using adaptive neuro fuzzy and hidden Markov model (HMM)	98.7%
Wenjie Zhang et al. [14]	Support vector machine (SVM)	Not mentioned
Huseyin Coskun et al. [15]	Artificial neural network (ANN)	90%
Sunjing et al. [16]	Convolutional neural network (CNN)	80.2%
Diogo Marcelo Nogueira1 et al. [17]	Support vector machine (SVM), CNN and random Forest	86.97%
Mohammad Nassralla et al. [18]	Neural network (NN): random forests, decision trees	92%
Wenjie Zhang et al. [19]	Support vector machine (SVM)	Not mentioned
Bradley M. Whitaker et al. [20]	Support vector machine (SVM)	56.4%
Lubaib P. et al. [21]	SVM classifier, KNN, Bayesian and Gaussian mixture model	99.99%
Shi-Wen Deng et al. [22]	Support vector machine (SVM)	74%
Heechang Ryu et al. [23]	Convolutional neural network (CNN)	79.5%
Edmund Kay et al. [24]	Drop connect classifier	85.2%
Tanachat Nilanon et al. [25]	Baseline classifier includes logistic regression, SVM and random forests and CNN	Not mentioned
Gari D. Clifford et al. [26]	Benchmark classifier	86.02%
Christian Thomae et al. [27]	ANN – Multi-layer perceptron	Not mentioned

(Continued)

Application of Soft Computing Techniques to Heart Sound Classification   **129**

TABLE 7.3 *(Continued)*

Reference	Classifiers Used	% of Accuracy
Bradley M. Whitaker et al. [28]	Support vector machine (SVM)	Not mentioned
Soo-Kng Teo et al. [29]	Neural network (NN)	76.7%
Michael Tschannen et al. [30]	Support vector machine (SVM)	81.2%
Mohamed Moustafa Azmy [31]	Support vector machine (SVM)	92.29%
Simarjot Kaur Randhawa et al. [32]	KNN, fuzzy KNN and ANN	99.6%
V. Nivitha Varghees et al. [33]	Decision rule classifier	82.76%–100%
Fatemah Safara [34]	Support vector machine (SVM)	99.39%
Shivnarayan Patidar et al. [35]	Least square-support vector machine (LS-SVM	94.01%
Laurentius Kuncoro Probo Saputra et al. [36]	Artificial neural network (ANN)	92.31%
Grzegorz Redlarski et al. [37]	SVM-MCS (modified cuckoo search)	93%
Fatemeh Safara et al. [38]	Support vector machine (SVM)	97.56%
Elsa Ferreira Gomes et al. [39]	Decision tree and multi-layer perceptron	72.76%
M. Markaki et al. [40]	Support vector machine (SVM)	96.07%
Elsa Ferreira Gomes et al. [41]	Multi-layer perceptron (MLP)	Not mentioned
C. Kwak et al. [42]	Support vector machine (SVM)	85.6%
Fatemeh Safara et al. [43]	Decision tree, KNN, Bayes Net, MLP and SVM	96.94% and 95.33%
Di Zhang et al. [44]	Linear discriminant analysis (LDA)	Not mentioned
Gur Emre Guraksin et al. [45]	Least square-support vector machine (LS-SVM)	96.66%
Samjin Choi et al. [46]	Multi-support vector machine (SVM)	Not mentioned
Guy Amit et al. [47]	K-nearest neighbour (KNN)	86.7%
Zumray Dokur et al. [48]	Kohonen's self-organising map (SOM) network and incremental self-organising map (ISOM)	95% and 75%

7.4 RESEARCH GAP

The study is limited to the review of 10 years of research papers from 2008 to 2018 on heart sound classification where different methods, approaches, techniques, algorithms, findings, result comparison and so on were studied. Table 7.4 describes the decade yearwise research gap of review papers.

TABLE 7.4

Research Gap of Various Review Papers

Reference	Year	Proposed Work/ Developed Model	Future Work
Bin Xiao et al. [5]	2018	For classification of heart sound, a novel 1-D CNN architecture is developed. Discriminative features are extracted. Also achieves state-of-art performance for classification.	Environment noises will take to build a model with more robustness. For mobile or embedded applications, some efficient architecture will be developed.
Shahid Ismail Malik et al. [3]	2018	Proficient strategy is utilised for confinement and grouping of heart thumps in commotion-based signals. For restriction window-based pinnacle handling, calculation is utilised, and for characterisation, history-based classifier is utilised	Time space and recurrence area highlights like zero intersections, unearthly motion can be utilised to upgrade the proposed procedure. Period-based preparing can be utilised in further research. Observational mode decomposition (EMD), wavelet can be utilised for signal decomposition.
Shanti R. Thiyagaraja et al. [6]	2018	Proposed an electronic stethoscope which can record or identify 16 different types of heart sound signals. The stethoscope is smart phone-based, which is portable and low in cost.	The model will improve to identify murmurs of heart and for other body sounds such as lungs.
Fatima Chakir et al. [7]	2018	Pascal classifying heart sound challenge; used new techniques and methods to improve performances. To optimise the processing of signals, different methodology for each dataset is developed.	Phonocardiogram and electrocardiogram can be merged for early detection of the heart abnormalities.

(Continued)

Application of Soft Computing Techniques to Heart Sound Classification 131

TABLE 7.4 *(Continued)*

Reference	Year	Proposed Work/ Developed Model	Future Work
Maryam Hamidi et al. [11]	2018	Proposed two methods: In first method, they found a function to estimate PCG signal. In the second method, the fractal dimension improved the classification result with MFCC. The result of proposed method is superior in comparison of different methods.	Collection of more clinical data of different types of diseases. New methods are also to be investigated. The main goal is the implementation of hardware in the system as a tool in the clinical environment.
Sunjing et al. [16]	2017	Convolutional neural network is used to classify the heart sound signals. Pre-treatment, envelope extraction, classification and recognition are the process of classifying heart sound.	Classification and identification of heart sound signals, convolutional neural network-based method can be used.
Diogo Marcelo Nogueira1 et al. [17]	2017	To classify the normal and abnormal heart sound, support vector machine radial basis algorithm is used. Approximately 86.97% accuracy is obtained. MFCC features are used for feature extraction.	There is a problem with identifying minority class, and the result shows that the dataset is unbalanced. To balance the dataset, more training data can be collected to improve the results. They will focus on using CNN in large datasets.
Heechang Ryu et al. [23]	2016	Proposed a cardiac diagnostic model using CNN that can predict the normal or abnormal heart sound by classifying phonocardiography (PCG) in clinical and nonclinical environment.	—

(Continued)

132 Soft Computing Applications and Techniques in Healthcare

TABLE 7.4 *(Continued)*
Research Gap of Various Review Papers

Reference	Year	Proposed Work/ Developed Model	Future Work
Edmund Kay et al. [24]	2016	An algorithm is produced which is capable of classifying normal and abnormal HS with 80% of accuracy.	An algorithm can be made which can recognise a good-quality signal. For specific pathologies diagnosing, they will try to develop an algorithm.
Tanachat Nilanon et al. [25]	2016	The aim of proposed work is to identify patients quickly for further expert diagnosis. Also focuses on normal and abnormal classification of HS automatically.	Improving performances of convolutional neural network model
Mohamed Moustafa Azmy [31]	2015	Presented a new mother wavelet method that is used for classification of HS signals. To get the features, heart sound statistical components are calculated. SVM classifier is used to classify heart sound signals.	Linear discriminant analysis (LDA) and neural network (NN) classifier will be used for classification. Suggested to use principal component analysis (PCA) to extract features of heart sound.
V. Nivitha Varghees et al. [33]	2015	Using stationary wavelet transform (SWT) and Hilbert phase envelope, a noise-robust method is used for detection and classification of heart murmurs. SWT-based PCG signal decomposition is used for isolating the low-frequency artifacts and recording instrument noises and discriminating the heart sound (S1, S2, S3 and S4) from heart murmurs.	—

(Continued)

Application of Soft Computing Techniques to Heart Sound Classification   133

TABLE 7.4 *(Continued)*

Reference	Year	Proposed Work/ Developed Model	Future Work
Elsa Ferreira Gomes et al. [39]	2013	SAX-based motif discovery used in time series which improves precision or recall of normal and abnormal class with the use of best peak-based model.	Exploring of second-order approach which will relate multiple motifs. Clustering of motifs will be used for smoothing differences between motifs. Motifs of motifs is used for identifying frequent motifs.
Elsa Ferreira Gomes et al. [41]	2012	Present a methodology for classifying of heart sound PASCAL challenge. For S1 and S2 heart sound identification, an algorithm is proposed. J48 and MLP is used to train and classify the features.	For further analysis and good performance of heart sound signals, there is a need to improve the criteria for identifying the S1 and S2 sounds.
Samjin Choi et al. [46]	2010	Two classification techniques are used for murmur of cardiac sound: AR spectral analysis and multi-SVM. Removes the artifact noises using WD. AR-PSQ estimation method is used to get the cardiac sound.	—
Guy Amit et al. [47]	2009	A framework is developed for analysis of HS; hierarchical clustering and classification algorithm is used.	For cardiovascular and cardiopulmonary diseases, a new technology can be used to detect and diagnosis of mechanical dysfunctions. Also will try to enhance the proposed framework.

7.5 CONCLUSION

Following is a summation of this chapter's contents:

- This chapter completed a systematic review of the methodologies, approaches, techniques, proposed work and future scope for heart sound classification of the last 10 years.
- It included a review of databases from where the data are collected for training and testing of various classifiers.

134 Soft Computing Applications and Techniques in Healthcare

- It was observed that support vector machine (SVM) is one of the classifiers which is mostly used by the authors and mel-frequency cepstral coefficients (MFCC) is used to extract features.
- A smartphone application was also developed by a few authors for the purpose.
- Some authors accepted the challenges of PhysioNet/CinC Challenge 2016 and PASCAL Classifying Heart Sound Challenge and found better accuracy results.
- It also listed the future proposed work reported by the authors.

Still it is a challenging job to classify heart sounds recordings. Therefore, it is an important area for future research where many new algorithms and techniques will be applied. This chapter will help the reviewers for further research in the area of classification of heart sounds.

REFERENCES

1. Shi-Wen Deng, and Ji-Qing Han, Towards heart sound classification without segmentation via autocorrelation feature and diffusion maps, Future Generation Computer System, Elsevier, ISSN: 0167-739X, Vol. 60, pp. 13–21, 2016.
2. 'Cardiovascular diseases (CVDs): Fact Sheet', World Health Organization, September 2016[Online], Available: http://www.who.int/mediacentre/factsheets/fs317/en/
3. Shahid Ismail Malik, Muhammad Usman Akram, and Imran Siddiqi, Localization and classification of heartbeats using robust adaptive algorithm, Biomedical Signal Processing and Control, Elsevier, ISSN: 1746-8094, Vol. 49, pp. 57–77, 2018.
4. Wikipedia (n.d.), 'Heart Sounds', https://en.wikipedia.org/wiki/Heart_sounds. Last modified 29 April, 2020.
5. Bin Xiao, Yunqiu Xu, Xiuli Bi, Junhui Zhang, and Xu Ma, Heart sound classification using a novel 1-D convolutional neural network with extremely low parameter consumption, Neurocomputing, Elsevier, ISSN: 0925-2312, Vol. 392, 153–159, 7 June 2020, Available: https://doi.org/10.1016/j.neucom.2018.09.101.
6. Shanti R. Thiyagaraja, Ram Dantu, Pradhumna L. Shrestha, Anurag Chitnis, Mark A. Thompson, Pruthvi T. Anumandla, Tom Sarma, and Siva Dantu, A novel heart-mobile interface for detection and classification of heart sounds, Biomedical Signal Processing and Control, Elsevier, ISSN: 1746-8094, Vol. 45, pp. 313–324, 2018.
7. Fatima Chakir, Abdelilah Jilbab, Chafik Nacir, and Ahmed Hammouch, Phonocardiogram signals processing approach for PASCAL classifying heart sounds challenge, signal, image and video processing, Springer, ISSN: 1863-1711, Springer-Verlag London Ltd., part of Springer Nature, 2018, Available: https://doi.org/10/1007/s11760-018-1261-5.
8. Baris Bozkurt, Ioannis Germanakis, and Yannis Stylianou, A study of time-frequency features for CNN-based automatic heart sound classification for pathology detection, Computers in Biology and Medicine, Elsevier, ISSN: S0010-4825, Vol. 100, 130–143, 2018, doi: 10.1016/j.compbiomed.2018.06.026.
9. Qurat-ul-Ain Mubarak, Muhammad Usman Akram, Arslan Shaukat, Farhan Hussain, Sajid Gul Khawaja, and Wasi Haider Butt, Analysis of PCG signals using quality assessment and homomorphic filters for localization and classification of heart sounds, Computer Methods and Programs in Biomedicine, Elsevier, ISSN: S0169-2607, Vol. 164, pp. 143–157, 2018, doi:10.1016/j.cmpb.2018.07.006.
10. Hong Tang, Ziyin Dai, Yuanlin Jiang, Ting Li, and Chengyu Liu, PCG Classification using Multidomain Features and SVM Classifier, BioMed Research International, Article ID 4205027, Vol. 2018, pp. 1–14, 2018, Available: https://doi.org/10.1155/2018/4205027.

Application of Soft Computing Techniques to Heart Sound Classification   135

11. Maryam Hamidi, Hassan Ghassemian, and Maryam Imani, Classification of heart sound signal using curve fitting and fractal dimension, Biomedical Signal Processing and Control, Elsevier, ISSN: 1746-8094, Vol. 39, pp. 351–359, 2018.

12. Kimitake Ohkawa, Masaru Yamashita, and Shoichi Matsunaga, Classification between Abnormal and Normal Respiration through Observation rate of Heart Sounds within Lung Sounds, 26th European Signal Processing Conference (EUSIPCO), ISBN: 978-90-827970-1-5, pp. 1142–1146, 2018.

13. H.M. Fahad, M. Usman Ghani Khan, Tanzila Saba, Amjad Rehman, and Sajid Iqbal, Microscopic abnormality classification of cardiac murmurs using ANFIS and HMM, Microscopy Research Technique, WILEY, Vol. 81, Issue 5, pp. 449–457, 2018.

14. Wenjie Zhang, Jiqing Han, and Shiwen Deng, Heart sound classification based on scaled spectrogram and tensor decomposition, Expert Systems with Applications, Elsevier, ISSN: 0957-4174, Vol. 84, pp. 220–231, 2017.

15. Huseyin Coskun, Omer Deperlioglu, and Tuncay Yigit, Classification of Extra systole Heart Sounds with MFCC features by using Artificial Neural Network, 25th Signal Processing and Communications Applications Conference (SIU), IEEE, ISBN: 978-1-5090-6494-6, 2017, doi: 10.1109/SIU.2017.7960252

16. Sunjing, Kang Liru, Wang Weilian, and Songshaoshuai, Heart Sound Signals based on CNN Classification Research, ICBBS '17 Proceedings of the 6th International Conference on Bioinformatics and Biomedical Science, ISBN: 978-1-4503-5222-2, pp. 44–48, 2017, doi: 10.1145/3121138.3121173.

17. Diogo Marcelo Nogueira, Carlos Abreu Ferreira, and Alfpio M. Jorge, Classifying Heart Sounds using images of MFCC and Temporal Features, EPIA Conference on Artificial Intelligence, Springer, Vol. 10423, pp. 186–203, pp. 2017.

18. Mohammad Nassralla, Zeina El Zein, and Hazem Hajj, Classification of Normal and Abnormal Heart Sounds, Fourth International Conference on Advances in Biomedical Engineering (ICABME), IEEE, ISSN: 2377-5696, 2017, doi: 10.1109/ICABME.2017.8167538

19. Wenjie Zhang, Jiqing Han, and Shiwen Deng, Heart sound classification based on scaled spectrogram and partial Least Squares Regression, Biomedical Signal Processing and Control, Elsevier, ISSN: 1746-8094, Vol. 32, pp. 20–28, 2017.

20. Bradley M. Whitaker, Pradyumna B. Suresha, Chengyu Liu, Gari D. Clifford, and David V. Anderson, 'Combining sparse coding and time-domain features for heart sound classification', Physiological Measurement, IOP Science, Vol. 38, Issue 8, pp.1701–1713, 2017.

21. P. Lubaib, and K. V. Ahammed Muneer, The Heart Defect Analysis Based on PCG Signals Using Pattern Recognition Techniques, International Conference on Emerging Trends in Engineering, Science and Technology (ICETEST) 2015, Science Direct, Elsevier, Procedia Technology, ISSN: 2212-0173, Vol. 24, pp. 1024–1031, 2016.

22. Shi-Wen Deng, and Ji-Qing Han, Towards heart sound classification without segmentation via autocorrelation feature and diffusion maps, Future Generation Computer System, Elsevier, ISSN: 0167-739X, Vol. 60, pp. 13–21, 2016.

23. Heechang Ryu, Jinkyoo Park, and Hayong Shin, Classification of Heart Sound Recordings Using Convolution Neural Network, Computing in Cardiology Conference (CinC), IEEE, ISSN 2325-887X, Vol. 43, 2016, doi: 10.22489/CinC.2016.329-134.

24. Edmund Kay, and Anurag Agarwal, Drop Connected Neural Network Trained with Diverse Features for Classifying Heart Sounds, Computing in Cardiology Conference (CinC), IEEE, ISSN: 2325-887X, Vol. 43, 2016, doi:10.22489/CinC.2016.181-266.

25. Tanachat Nilanon, Jiayu Yao, Junheng Hao, Sanjay Purushotham, and Yan Liu, Normal/Abnormal Heart Sound Recordings Classification Using Convolutional Neural Network, Computing in Cardiology Conference (CinC), IEEE, ISSN: 2325-887X, Vol. 43, 2016, doi:10.22489/CinC.2016.169-535.

26. Gari D. Clifford, Chengyu Liu, Benjamin Moody, David Springer, Ikaro Silva, Qiao Li, and Roger G. Mark, Classification of Normal/Abnormal Heart Sound Recordings: the PhysioNet/Computing in Cardiology Challenge 2016, Computing in Cardiology Conference (CinC), IEEE, ISSN: 2325-887X, Vol. 43, 2016, doi:10.22489/CinC.2016.179-154.
27. Christian Thomae, and Andreas Dominik, Using Deep Gated RNN with a Convolutional Front End for End-to-End Classification of Heart Sound, Computing in Cardiology Conference (CinC), IEEE, ISSN: 2325-887X, Vol. 43, 2016, doi:10.22489/CinC.2016.183-214.
28. Bradley M. Whitaker, and David V. Anderson, Heart Sound Classification via Sparse Coding, Computing in Cardiology Conference (CinC), IEEE, ISSN: 2325-887X, Vol. 43, 2016, doi:10.22489/CinC.2016.234-191.
29. Soo-Kng Teo, Bo Yang, Ling Feng, and Yi Su, Power Spectrum Analysis for Classification of Heart Sound Recording, Computing in Cardiology Conference (CinC), IEEE, ISSN: 2325-887X, Vol. 43, 2016, doi:10.22489/CinC.2016.340-235.
30. Michael Tschannen, Thomas Kramer, Gian Marti, Matthias Heinzmann, and Thomas Wiatowski, Heart Sound Classification Using Deep Structured Features, Computing in Cardiology Conference (CinC), IEEE, ISSN: 2325-887X, Vol. 43, 2016, doi:10.22489/CinC.2016.162-186.
31. Mohamed Maoustafa Azmy, Classification of Normal and Abnormal Heart Sounds using New Mother Wavelet and Support Vector Machines, 4th International Conference on Electrical Engineering (ICEE), IEEE, ISBN: 978-1-4673-6673-1, 2015, doi: 10.1109/INTEE.2015.7416684.
32. Simarjot Kaur Randhawa, and Mandeep Singh, Classification of Heart Sound Signals using Multi-Modal Features, Procedia Computer Science, Elsevier, ISSN: 1877-0509, Vol. 58, pp. 165–171, 2015.
33. V. Nivitha Varghees, and K. I. Ramachandran, Heart Murmur Detection and Classification using Wavelet Transform and Hilbert Phase Envelope, Twenty First National Conference on Communications (NCC), IEEE, ISBN: 978-1-4799-6619-6, 2015, doi: 10.1109/NCC.2015.7084904
34. Fatemeh Safara, Cumulant-based trapezoidal basis selection for heart sound classification, Medical & Biological Engineering & Computing, Springer, Volume 53, Issue 11, pp. 1153–1164, 2015.
35. Shivnarayan Patidar, and Ram Bilas Pachori, Classification of cardiac sound signals using constrained tunable-Q wavelet transform, Expert system with Applications, Elsevier, ISSN: 0957-4174, Vol. 41, pp. 7161–7170, 2014.
36. Laurentius Kuncoro Probo Saputra, Hanung Adi Nugroho, and Meirista Wulandari, Feature Extraction and Classification of Heart Sound based on Autoregressive Power Spectral Density (AR-PSD), The 1st International Conference on Information Technology, Computer, and Electrical Engineering, IEEE, ISBN: 978-1-4799-6432-1, 2014, doi: 10.1109/ICITACEE.2014.7065730.
37. Grzegorz Rediarski, Dawid Gradolevski, and Aleksander Palkowski, A system for heart sounds classification, PLoS ONE, Vol. 9, Issue 11, e112673, 2014, Available: https://doi.org/10.1371/ journal.pone.0112673
38. Fatemah Safara, Shyamala Doraisamy, Azreen Azman, Azrul Jantan, Asri Ranga, and Abdullah Ramaiah, Multi-level basis selection of wavelet packet decomposition tree for heart sound classification, Computers in Biology and Medicine, Elsevier, ISSN: 0010-4825, Vol. 43, pp. 1407–1414, 2013.
39. Elsa Ferreira Gomes, Allpio M. Jorge, and Paulo J. Azevedo, Classifying Heart Sounds using Multiresolution Time Series Motifs: an exploratory study, Proceedings of the International C* Conference on Computer Science and Software Engineering, ISBN: 978-1-4503-1976-8, pp. 23–30, 2013, doi: 10.1145/2494444.2494458

40. M. Markaki, I. Germanakis, and Y. Stylianou, Automatic Classification of Systolic Heart Murmurs, IEEE International Conference on Acoustics, Speech and Signal Processing, IEEE, ISBN: 978-1-4799-0356-6, pp. 1301–1305, 2013, doi: 10.1109/ICASSP.2013.6637861

41. Elsa Ferreira Gomes, and Emanuel Pereira, Classifying heart sounds using peak location for segmentation and feature construction, Manuscript under review by AISTATS, 2012.

42. C. Kwak, and O.W. Kwon, Cardiac disorder classification by heart sound signals using murmur likelihood and hidden Markov model state likelihood, IET Signal Processing, Vol. 6, Issue 4, pp. 326–334, 2012, doi:10.1049/iet-spr.2011.0170.

43. Fatemah Safara, Shyamala Doraisamy, Azreen Azman, Azrul Jantan, Asri Ranga, and Abdullah Ramaiah, Wavelet packet entropy for heart murmurs classification, Advances in Bioinformatics, Hindawi, Article ID 327269, Vol. 2012, pp. 1–6, 2012, Available: http://dx.doi.org/10.1155/2012/327269

44. Di Zhang, Jiazhong He, Yueping Jiang, and Minghui Du, Analysis and classification of heart sounds with mechanical prosthetic heart valves based on Hilbert-Huang transform, International Journal of Cardiology, Elsevier, Vol. 151, Issue 1, pp. 126–127, 2011, Available: https://doi.org/10.1016/j.ijcard.2011.06.033

45. Gur Emre Guraksin, and Harun Uguz, Classification of heart sounds based on the least squares support vector machine, International Journal of Innovative Computing, Information and Control, ISSN: 1349-4198, Vol. 7, Issue 12, pp. 7131–7144, 2011.

46. Samjin Choi, and Zhongwei Jiang, Cardiac sound murmurs classification with autoregressive spectral analysis and multi-support vector machine technique, Computers in Biology and Medicine, Elsevier, ISSN: 0010-4825, Vol. 40, pp. 8–20, 2010.

47. Guy Amit, Noam Gavriely, and Nathan Intrator, Cluster analysis and classification of Heart Sounds, Biomedical Signal Processing and Control, Elsevier, ISSN: 1746-8094, Vol. 4, pp. 26–36, 2008.

48. Zumray Dokur, and Tamer Olmez, Heart sound classification using wavelet transform and incremental self-organizing map, Digital Signal Processing, Elsevier, ISSN: 1051-2004, Vol. 18, pp. 951–959, 2008.

8 Fuzzy Systems in Medicine and Healthcare
Need, Challenges and Applications

Deepak K. Sharma[1], Sakshi[2] and Kartik Singhal[2]
[1]Department of Information Technology
Netaji Subhas University of Technology
New Delhi, India
[2]Department of Manufacturing Process
  and Automation Engineering
Netaji Subhas University of Technology
New Delhi, India

CONTENTS

8.1 Introduction 140
    8.1.1 Introduction to Fuzzy Logic Systems 140
    8.1.2 Fuzzy Set Theory.. 141
    8.1.3 Applications of Fuzzy Systems in Healthcare 141
8.2 Fuzzy Systems in Healthcare... 142
    8.2.1 Challenges to Healthcare 142
    8.2.2 Soft Computing Techniques Involved with
        Fuzzy Systems 143
        8.2.2.1 Fuzzy Cognitive Maps 144
        8.2.2.2 Neuro-Fuzzy Systems 145
        8.2.2.3 Genetic Algorithms.. 146
        8.2.2.4 Bayesian Networks... 147
    8.2.3 Fuzzy Logic in Medical Information Management......................... 147
        8.2.3.1 Medical Databases 147
        8.2.3.2 Information Retrieval.. 148
8.3 Role of Fuzzy Systems in Diagnosis and Risk to Health Analysis 148
    8.3.1 Lung Cancer 148
    8.3.2 Heterogeneous Childhood Cancers 149
        8.3.2.1 Breast Cancer... 149
        8.3.2.2 Diagnosis of Diseases 149

	8.3.2.3	Periodontal Dental Disease	149
	8.3.2.4	Anaemia	150
	8.3.2.5	HIV	151
	8.3.2.6	Malaria	151
	8.3.2.7	Diabetes	152
	8.3.2.8	Asthma	153
	8.3.3	Monitoring Patient's Condition during Heart Surgery	153
8.4	Role of Fuzzy Systems in Medicine		153
	8.4.1	Analysis of Signs or Symptoms	154
	8.4.2	Medical Image Processing	155
	8.4.3	Administration of Anaesthesia	155
	8.4.4	Determination of Drug Dose	156
	8.4.5	Improving Service Quality	156
	8.4.6	Service Quality Evaluation in Yoncali Physiotherapy and Rehabilitation Hospital—Case Study	157
8.5	Conclusion		158
References			159

8.1 INTRODUCTION

8.1.1 INTRODUCTION TO FUZZY LOGIC SYSTEMS

Healthcare and medicine are fields of ambiguity and a multitude of possibilities. Each diagnosis or prognosis made can be affected by a plethora of factors, many unknown, and even a minute change reported could lead to a distinct interpretation. Thus, contrary to the stiff nature of hard computing techniques, soft computing methods such as fuzzy logic are required to incorporate *fuzziness* in health. Fuzzy logic is an extension of the domain of two-value logic, as it deals with the states that lie in between the extreme notations of true or false. It was first established in 1965 by Lotfi Zadeh on account of his work in mathematics of fuzzy sets [1]. Since then, it has been widely accepted in areas where approximate reasoning is inherent. Instead of dealing in numbers or binary logic to develop computational intelligence-based systems, fuzzy logic provides the capability to communicate in linguistic terms with the computer.

This nature of fuzzy logic systems makes them particularly suitable in dealing with non-crisp or imprecise data and the study of rule-based systems [2]. It also provides better tolerance to approximations, vagueness or uncertainty in data. An effective fuzzy logic system requires structural planning, expert knowledge and a validation process. For any fuzzy system, it is imperative to first realise all the required input factors, develop a rule base of *if-then* rules, define the necessary membership functions and model parameters. The set of rules is generally in accordance with the knowledge base and relevant conditions for the system, designed under expert supervision [3]. Finally, the proposed system is tuned in accordance with the consequences produced.

The process of designing fuzzy systems is review-intensive, as the assignment of membership functions, model parameters and the ruleset is based on instinct and knowledge of the expert. Thus, validation procedure for the system becomes an important aspect. Fortunately, the procedure is highly optimised with the help of

Fuzzy Systems in Medicine and Healthcare 141

other soft computing techniques such as Bayesian inferences, neural networks, particle swarm optimisation methods or evolutionary computation methods like genetic algorithms [4]. The scope of their applications and methodology will be discussed in detail further in this chapter.

8.1.2 FUZZY SET THEORY

Fuzzy set theory is used in defining problem statements which are not completely understandable within the scope of the two-value logic system. Consider the following statement: *'Is this water suitable for drinking?'* Now, in the two-value logic system, we can follow the convention of contaminated being 0 value (False) and clean being 1 (True). A system based on the above-mentioned notion will only provide two outputs: contaminated or clean. But it will fail to address the level of contamination or pureness. Meanwhile, a fuzzy logic system will weigh in the level of contamination to provide the output as maybe: *'somewhat clean or highly contaminated'*. It is a very common misunderstanding to confuse fuzzy set theory with probabilistic theory due to the inclusion of terms like the degree of truth or association of value. But the reality is both are very different in their approach. Now consider this statement: *'This glass is 64% contaminated'*, if we follow probabilistic theory, the above statement will mean there is a 64% chance the water in the glass is contaminated. But by fuzzy set theory, it will simply signify the level of contamination.

A fuzzy system *maps* the input value to the expert knowledge base using the predefined *if-then* rule set, and then presents a consequence, i.e., the output value as the result of the best matching proposition. The data is firstly assigned a *degree of belongingness* to the most appropriate set under knowledge base [5]. It is then fuzzified—that is, converted into granular (linguistic) terms for further computation. Then this granular data is processed in accordance with the rules of the system and a fuzzified output is produced. Lastly, the fuzzified output is defuzzified and is presented as a crisp value. The *belongingness* of the fuzzy data is characterised by the use of membership functions. In classic set theory, a value 'x' is either an element of the set A, if, $x \in$ A or it is not, if, $x \notin$ A. In fuzzy set theory, the input numerical value can also be associated partially with the membership functions within the *universe of discourse*—that is, it can be a member of Set A even with only *some degree of truth*. It should be noted that the *degree of membership* is also the weight of the association of value to the respective fuzzy set [6]. The closer it is to 1, the higher is the *belongingness* to the particular set.

8.1.3 APPLICATIONS OF FUZZY SYSTEMS IN HEALTHCARE

Healthcare and medicine is a field of acute uncertainty [7]. Be it knowledge acquisition, data management or administration of drug dosage, use of linguistic variables is necessary to improve understanding in human-friendly terms. The data associated with a patient's health or a diagnosis can be very easily classified as fuzzy. A patient may define his fever as 'high' or 'low'; a diagnosis may show 'rapid' or 'lax' tumour growth. Similarly, almost all the data available in the sector is imprecise in nature, thus making fuzzy logic systems one of the most effective tools for the industry [8].

142   Soft Computing Applications and Techniques in Healthcare

Fuzzy systems are used in various applications throughout the industry, both as standalone features as well as in amalgamation with other soft computing techniques such as in advanced areas of monitoring tumour growth or classification of cancer cells. They also provide an effective choice for maintaining and analysing patient records, as data from multiple diagnoses or sources can often be confusing.

Researchers and practitioners have utilised fuzzy logic as a manner of assigning any computer to undertake a task meant for a human operator [9]. As established earlier in this chapter, healthcare and medicine has become one of the most innovative application areas for fuzzy logic, although the process of developing efficient systems is still dependent on operator knowledge and available relevant data. By virtue, the most accepted practices become the standard in the development of computer-based intelligent systems. Many cases still exist where no previous algorithm or available system could be redesigned as a fuzzy expert system. Thus, a fuzzy logic-based system in such a case is only as useful as the capability of the expert who supervised the development of its fuzzy sets and inference rules. The proficiency of a system is higher if the said individual here is considered an authority or an expert in the area, but still it in no way guarantees 100% success.

Numerous approaches and methodologies have been studied over the years to deal with the variations that arise in uncertain situations and specifically in the domain of medicine. The fuzzy systems that are modelled are based on the available knowledge and through the process of medical practitioners. Novel methodologies include the fields of medical knowledge acquisition, hybrid systems, treatment planning, decision making and advisory, monitoring and control, unknown parameter predictions or optimisations, artificial thinking and others. Fuzzy models have been widely utilised in the design of support systems based on the experience of the expert's knowledge in the clinical context. The studies presented in this chapter are intended to discuss existing conceptual models of fuzzy logic-based systems and explore their utility in healthcare and medicine. Data interpretation, pattern recognition, segmentation and classification of data using the fuzzy approach are discussed. Along with fuzzy logic, other computing methods such as neural networks or genetic algorithms are combined to build hybrid systems for diagnosis purposes.

This chapter will first address the current situation of the healthcare sector and the role of fuzzy systems. It will then focus on various soft computing techniques used with fuzzy systems and study their operations and development. In later parts of this chapter, the contribution of fuzzy systems in the cure of many diseases will be studied, followed by a discussion on the role of fuzzy logic-based controllers in supporting medical decision-making processes. This chapter concludes by exploring the developments and progress of fuzzy systems in recent years and their future scope within the domain of healthcare and medicine.

8.2 FUZZY SYSTEMS IN HEALTHCARE

8.2.1 CHALLENGES TO HEALTHCARE

Quality healthcare forms an integral part of an individual's perception of life. The change in lifestyle and emergence of newer practices is introducing new challenges

Fuzzy Systems in Medicine and Healthcare

or bringing the old ones into the limelight [10]. The scope to revolutionise the way care is delivered can be seen as an opportunity among the most progressive health economies. Some of the impediments that will impact the medical industry in the future include a rise in the costs, fulfilling the demand for service and improving the quality of care.

Records indicate that the current rate at which healthcare spending is growing has already surpassed the growth levels of GDP. The receivers and the providers of healthcare are faced with challenging macroeconomic factors like the older population or lack of public funding. Studies show that adoption rates of medical information systems vary greatly from country to country. In fact, the consumers are dependent on the structure of the healthcare system and not necessarily on the demographics. Additionally, recent trends indicate that there has been a trend of the growing demand for personal healthcare delivery organisations worldwide. With advancements in the identification and developing an understanding of the causes of diseases, life expectancy has proven to increase. This chapter further explains the importance of fuzzy logic-based systems and techniques to achieve an improved diagnosis of diseases. The demand for transparency of data and processes has encouraged responsible organisations to invest in improved and quality end products.

Healthcare, being one of the major supply-driven industries, will soon become an industry where people start demanding its transition to an industry that caters to their needs, fears and aspirations [11].

8.2.2 Soft Computing Techniques Involved with Fuzzy Systems

Healthcare is stochastic in nature. Instead of dealing in exact numbers and precise figures, practitioners deal with data in vague linguistic terms. Medical statements regarding the severity of a certain disease or nature of cancer cells cannot be assessed within the scope of binary logic or within precise numerical parameters. Subjectivity exists in the analysis of data and the decision-making procedure, which makes the final outcome highly dependent upon the knowledge of the human operator and the conditions at hand. The diagnosis procedure is very dependent on how a patient describes his or her condition [12]. It's very common for someone to describe his or her condition as 'not good', 'feeling unwell' or just 'better than the last clinic test'. All these 'details' may not provide an accurate representation of the actual condition, and it is plausible that a doctor or a practitioner makes an incorrect conclusion on the basis of initial tests or misleading description by the patient. Thus the experience of human operator and precision of supporting decision-making apparatus is also a major factor. For this reason, developing precise logical systems is both crucial and a tedious procedure. It has been a long-standing goal of the medical industry to minimise misinterpretation of clinical tests or patient's data and avoid errors in a medical procedure. Hopefully, improved soft computing techniques and access to better analytics of medical data show optimistic developments in the design of logical systems. This section will introduce different soft computing techniques used in improvising capabilities of decision support systems and popular methodology for the development of hybrid fuzzy systems.

8.2.2.1 Fuzzy Cognitive Maps

Fuzzy cognitive maps (FCMs) provide an illustrative representation of a fuzzy decision-making process [13]. An FCM is a network of nodes, also termed concepts, which are interconnected with links that are used to determine casual representations of the intrarelationships between the data. The links allow a forward-backwards propagation of the causality between the nodes. This structure is efficient in dealing with data that has intradependent parameters. Expert behaviour, model parameters and weights of each relationship are used in characterising an FCM structure. It is crucial to have knowledge of possible interactions between the various concepts and the effect of one's behaviour on others. An expert is required to suggest the linkage linguistic variables which are part of a fuzzy set to map different concepts and assign their degree of causality. Thus, fuzzy cognitive maps is an expert opinion-based technique which requires sufficient improvisation in order to exploit the available data and knowledge to develop an optimum behavioural system which can effectively learn from its experience.

In healthcare practices, FCMs are employed to support the human decision-making process in the presence of missing or even conflicting data or patterns [14]. FCMs have been successfully modelled to distinguish between different medical conditions when symptoms for more than one ailment are reported positive and in the development of predictive tools for identification of infectious diseases or in the administration of drugs. One more important utilisation of FCMs is in risk assessment systems or behaviour analysis studies. FCMs have provided positive results in enhancing strategy models for diagnosis or prognosis of vector-borne diseases (see Figure 8.1).

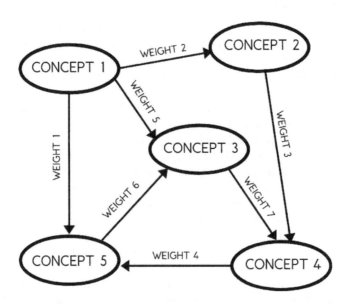

FIGURE 8.1 A fuzzy cognitive map (FCM) representing relations between different concepts. Each arrowhead shows the direction of the node and carries the weight of the dependency.

8.2.2.2 Neuro-Fuzzy Systems

Hybrid neuro-fuzzy systems or adaptive neuro-fuzzy inference systems (ANFIS) are valuable in the development of medical decision support systems (MDSS) and medical data analytics [15]. Designing a precise or effective fuzzy rule base is a challenging process. As mentioned earlier in this chapter, many soft techniques such as artificial neural networks are utilised with fuzzy systems to improve the process. Integration of neural networks with fuzzy systems provides an intuitive way in extracting rules for the inference engine and expanding the limits of neural networks' decision-making capabilities. The neural network makes up learning the core of the system and is generally a multilayered structure. In the design of ANFIS, an initial system including the relevant membership functions is developed and the input-output parameters are assigned. The designed system is then trained with designated datasets to improve the consequent parameters and improve the precision of the system. The most commonly used training methods are gradient descent or the least squares method.

Remarkable research has been done to contribute to the progress of ANFIS based on MDDSs. ANFIS provides a robust and resource-effective methodology in the prediction of many health conditions including Alzheimer's disease, osteoarthritis disease, Parkinson's disease, heart failure or even covert ailments such as Huntington's disease. Andrius Laurinaitis et al. introduced a novel method to study the symptoms of neurological disorders causing involuntary movements in patients of Huntington's disease [16]. They studied the reaction condition from finger tapping tests from HD patients and analysed the data using fuzzy logic systems (FLSs) for reaction state and ANN for prediction of patients' functional capacity levels. Ali Yadollahpour et al. discussed the development of a predictive system for progression of chronic kidney disease (CKD) [17]. Their study utilised data from CKD patients to model a Takagi-Sugeno type ANFIS for early detection of renal failure. Hybrid neuro-fuzzy systems have been utilised effectively in improving the accuracy of diagnosis. For example, ANFIS-based MDSS has reported up to 97% success rate in the identification of tuberculosis and 95.55% accuracy in the detection of lung cancer (see Figure 8.2) [18].

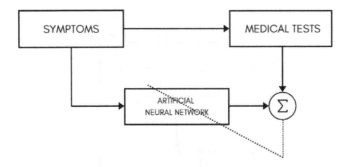

FIGURE 8.2 An artificial neural network system representation for diagnosis. The system is trained using past known cases.

8.2.2.3 Genetic Algorithms

Genetic algorithms are evolutionary computational techniques modelled to imitate the natural selection process. Like the biological evolution process, it imitates search and optimisation in the following steps: identification and selection of the population, evaluation of fitness grade and creation of population by introducing crossover or mutations. The process is based on the endurance of the fittest population, in which a set of data is selected as the initial population and future species or solutions are produced with desired characteristics or under required conditions [19]. The newly formed solution set or *offspring data* is used as the parent class for the next generation and so on. Desired characteristics can be introduced or primitive characteristics can be omitted from newer generations by inducing mutations or crossovers. Genetic algorithms are extensively used in optimising weights of rules in ANNs or fuzzy inference engines. It should be noted that it is a metaheuristic method and an iterative process.

Genetic algorithms are utilised to drive a solution towards an *optimum generation*. Medical history databases and patient's reports can be unreliable, as the data collected might be simply substandard for research purposes or scarce in the information. GAs are integrated with fuzzy standard additive models to provide an improvised data classification methodology. The GSAM computations developed are a low-cost and high-efficiency process. GSAMs are used extensively in data classification for breast cancer and heart disease patients. Further, GAs are also involved in tuning fuzzy systems for computing membership functions or improving consequential parameters. Rana Akhoondi, Rahil Hosseini and Mahdi Mazinani provided a GA-fuzzy-based approach to improve heart disease prediction rate from 85.52% to 92.37%, based on medical data from 380 patients in Iran (see Figure 8.3).

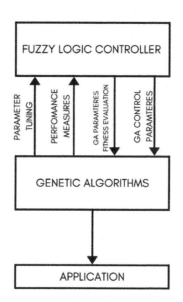

FIGURE 8.3 Flowchart depicting an FLC- and genetic algorithm-based system.

Fuzzy Systems in Medicine and Healthcare

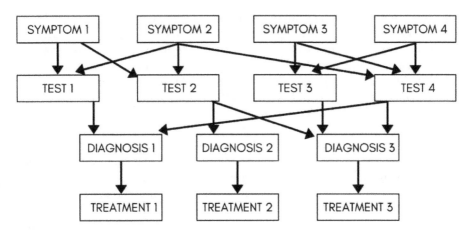

FIGURE 8.4 A Bayesian network representing a medical decision-making system in the presence of multiple symptoms.

8.2.2.4 Bayesian Networks

Bayesian networks are used in modelling risk assessment during medical procedures or surgeries. They are based on Bayesian inference for evaluating the causes by interpreting the relationships between the associated variables. The very basic step in designing a Bayesian network is to determine the aim or the scope of the system. This can be the desired objective, for example, calculating the success rate of heart surgery. Secondly, it is necessary to identify the pertinent constraints or parameters that can affect the objective. Expert knowledge and patient's data can be utilised in developing a risk factor model. Once the essential information is acquired, a Bayesian network is designed to represent the knowledge in graphical form with relevant conditional dependency as edges and variables as nodes. This is followed by assigning weights to each decision-making parameter or informational nodes followed by establishing qualitative reasoning between different nodes through conditional probabilities table. The last step is the analysis of the Bayesian network to understand the result. In a case of availability of data, Bayesian networks provide a well-organised platform for reasoning and analysis.

In case of non-availability of data or conflict in expert opinion, fuzzy systems can be integrated with Bayesian networks [20]. Fuzzy membership functions can be used to aggregate different expert opinions, and fuzzification of data can be done by incorporating qualitative vague linguistic terms. In this manner, fuzzification provides flexibility in risk assessment by dealing with the vague data or absence of certain events. Thus, fuzzy Bayesian networks provide a substantial proposition by replacing the need for quality data with an evaluation of subjective opinions or historical knowledge (see Figure 8.4).

8.2.3 FUZZY LOGIC IN MEDICAL INFORMATION MANAGEMENT

8.2.3.1 Medical Databases

The analysis of the huge amount of data present in medical databases requires the use of efficient data mining techniques in order to warehouse all these databases.

One of the most efficient data mining techniques includes the use of rough set theory (RST) and fuzzy logic. It is completed in two stages: clustering and classification. The first step is to use the rough set theory for the clustering process. The grouping of data that have the same characteristics into the same cluster is known as clustering. After the data undergoes this process, similar objects will be clustered together and dissimilar objects will be clustered together. RST is used to reduce the complexity associated with this process. The second phase works on the classification of these clusters by using fuzzy logic. A large number of rules are generated for classification. The complexity of the classification process is greatly reduced by the use of RST. Factors such as sensitivity, specificity and accuracy with different cluster numbers are used to determine the performance analysis [21].

8.2.3.2 Information Retrieval

Despite the attempts made by information retrieval systems to manage the integrity of the data collected through search engines and records, there are still many problems such as weak concepts representation, inaccuracy of retrieved information, and vague retrieval systems. Any fuzzy model contains an expert-supervised inference engine and fuzzy memberships which provide effective adaptive information retrieval along with an efficient representation for text. During the main process, the use of keyword and keyphrase extraction helps in the representation of the document. A user model, based on inferring fuzzy user modelling aspects, is used to represent the acquired user model in an ontological format. It provides many advantages such as flexibility and reliability, improving upon traditional problems encountered with search engines, and minimising the information retrieval process time that is expected from a potential system. The use of critical assessment metrics, including precision, recall, fall-out and f-measure, improve the accuracy of retrieved information rather than the traditional ways.

8.3 ROLE OF FUZZY SYSTEMS IN DIAGNOSIS AND RISK TO HEALTH ANALYSIS

Thousands of lives are lost to cancer due to the limitations of medical resources and the inability to efficiently inhibit its progress. Cancer is developed with an abnormal increase in cell population and anomaly in cell growth. Damage to DNA and cells is one of the major causes of this disease. Cancer diagnosed at early stages can be successfully treated.

8.3.1 LUNG CANCER

Lung cancer is the second most common type of cancer in terms of its occurrence. It is a significant contributing cause of death in women and men of all age groups, The best way to protect from this dangerous disease is to detect it early. One of the most promising ways to detect this disease at an early stage is the lung cancer diagnosis system based on fuzzy logic system and neural network. In this system, several techniques are used for improving the lung image and segmenting

Fuzzy Systems in Medicine and Healthcare 149

it to retrieve more knowledge about the characteristics of the CT lung image. This is followed by the use of the GLCM method to extract the features of the image which are used to classify the stage. Neural networks are used in the classification of the image [22]. The combined outcome, of the result of the neural network and a fuzzy system, is used to distinguish between the stages of cancer, based on the symptoms of the patient.

8.3.2 HETEROGENEOUS CHILDHOOD CANCERS

There are four heterogeneous cancers: neuroblastoma, non-Hodgkin lymphoma, rhabdomyosarcoma and Ewing sarcoma. The small round blue cell tumour (SRBCT) exhibited by these cancers often leads to errors in diagnosis. The small set of biomarkers are identified with the aid of multilayer networks with online gene selection ability and relational fuzzy clustering.

8.3.2.1 Breast Cancer

Breast cancer is one of the most fatal health problems faced by women. The timely identification of patients and the reduction in treatment expenses have been achieved through breast cancer screening.

The processing of data derived from patients with breast cancer is done to provide a risk prognosis to medical institutions. This prediction of the risk of breast cancer can be done by using a fuzzy logic technique. This technique utilises factors such as patient age and automatically extracted tumour features. The process utilises a thermal camera for imaging of the patients. The posterior analysis is done for the important parameters derived from the images. This is done with the aid of a genetic algorithm [23]. This is followed by the identification process of principal components of a fuzzy neural network for clustering breast cancer. The usage of non-radiation rays in thermal image scanning has helped it achieve preferential status in screening for women.

8.3.2.2 Diagnosis of Diseases

Over the past decade, most researchers have advocated for the increased application of computer-aided disease diagnosis methods so as to better decision-making capacity of health professionals. To date, various computational methods simulating the thinking process of experts have already been introduced and worked upon in this regard. Mapping a numeric result in linguistic expression with the help of fuzzy logic is done for comprehensive and improved decision making. The development of various systems, algorithms and applications by utilizing fuzzy logic shows that it is a significant approach in the diagnosis of chronic diseases.

8.3.2.3 Periodontal Dental Disease

Periodontal dental disease is a chronic diseases where its spreading in dental supporting tissues results in tooth loss [24].

Fuzzy expert system (FES) methods are known to yield better diagnosis and treatment results than traditional methods. The decision of the fuzzy system design is followed by the formulation of the 'if-then' rules. Using the Mandini approach as

150 Soft Computing Applications and Techniques in Healthcare

an output mechanism and Mamdani max-min rule to formulate the validity degree for each rule, the input variables in the designed FES are as follows:

- Gingival index: Ranging from 0–3 gi, its application lies in the examination of gingival health.
- Alveolar bone loss: Ranging between 0–9mm, its application lies in measuring bone loss.
- Probing pocket depth: Ranging between 0–8mm, its application lies in measuring tissue loss.
- Mobility(MB): Ranging between 0–3mm, its application lies in measuring dental activity.
- Attachment loss (AL): Ranging between 0–9 units, its application lies in measuring the fibre loss connecting gingival to bone

The certainty of periodontal disease includes gingivitis and periodontitis. Linguistic expressions of output variables were used for the formation of output values. This was followed by the formation of an FES using risk factors such as age along with these output values. The designing of the system was done by using a method similar to the examination form used in the treatment of disease. The system could determine the severity of the disease, after loading the relevant findings to FES. The output of the ES developed by the question-answer method suggested a suitable treatment method. The stage of the disease is indicated in the output according to the inputs of Gingival Index(GI), Mobility(MOB), Probing Pocket Depth(PPD), Attachment Loss(AL) and Alveolar Bone Loss(ABL) yield the stage of the disease as the output.

8.3.2.4 Anaemia

The deterministic systems for child anaemia have been designed on the basis of Takagi–Sugeno (TS) type fuzzy model consisting of the multi-input single-output neuro-fuzzy network. It consists of a five-layered architecture. The first and second input parameters are haemoglobin and haematocrit values, respectively. The use of triangular membership functions allows the easy gathering of their mathematical formulas.

The first step in the analysis includes the measurement of haemoglobin, haematocrit and anaemia levels. This is followed by the second layer which includes fuzzification, after which values are obtained for haemoglobin and haematocrit. In the rule layer or the third layer, two out of the nine rules are referenced. The next layer or the fourth layer includes the normalisation of the result followed by the defuzzification in the fifth layer. After the obtaining of the anaemic levels by the system, an FES was designed for the determination of the level of iron deficiency anaemia. Further, the centroid defuzzification method and the Mamdani inference mechanism were used together. The utilisation of the programming language as the result of the operations of the system helps to display the severity in an appreciable way. The input parameters specified in the system are Hb (haemoglobin) amount, MCV (mean corpuscular volume), SI (iron level in serum), TIBC (total iron-binding capacity) and the amount of ferritin. The level of IDA (iron deficiency anaemia) is the output parameter. Thus, 432 rules are reduced to 255.

Fuzzy Systems in Medicine and Healthcare

8.3.2.5 HIV

Constant efforts have been made at the national and provincial levels to develop effective schemes for preventing and controlling HIV/AIDS. The scope to decrease the levels of mortality and morbidity can be achieved by early detection of HIV. Many of the diagnostic tests which are already in place have proven to be extremely complex and subjective, depending largely on the expertise of the clinician.

The combined use of the neural network (NN) and fuzzy inference systems (neuro-fuzzy) for detecting the risk levels of patients with HIV has been proposed in this chapter. Due to its ability to carry out the user-friendly and accurate method of HIV diagnosis, using the neuro-fuzzy system, it has become one of the plausible ways to deal with the discrepancies of data. The diagnosis procedure of HIV is initiated when somebody presents certain complaints to a physician. The physician then gathers relevant information on the basis of a patient's response or clinical tests. Afterwards, the physician compiles a list of the conditions that could represent the patient's situation in its entirety with which the possibilities of illness are then narrowed down. The learning procedure of NN and the appropriate human reasoning capabilities of FL are then combined with the neuro-fuzzy inference procedure. The individual limitations of the two soft computing techniques mentioned above are overcome for producing an accurate medical report, which is then applied to the diagnosis procedure of HIV. The use of backpropagation helps in the training of the probable systems by the above-mentioned expert system comprised of a neural network for the same. The adjustment between the weight of the input layer and the output layer is made. The results are then fed into a fuzzy logic knowledge base. The linguistic variables put to use are low, moderate, and high and very high. These linguistic variables help in the generation of the knowledge base rules as the main output of the fuzzy system. These rules can be presented in the form of **IF x is a THEN y is b.** The input/output values applied to the rules are mapped with membership functions. For defuzzification of the fuzzified input into a crisp value, the root mean square (RMS) method is used.

8.3.2.6 Malaria

Malaria remains one of the world's greatest problems in terms of morbidity and mortality. Treatment and diagnosis of malaria is a difficult task, especially for regions with low resources or lack of medical personnel. Thus, employing techniques like fuzzy logic in the treatment of such diseases has been a keen interest of researchers. Eventually, a good amount of advanced fuzzy expert systems have been designed in the diagnosis and treatment of malaria (see Figure 8.5). An example of the algorithm of the FES for the diagnosis is as follows:

1. Feeding signs and symptoms related to the patient complaint, where m represents the number of signs and symptoms.
2. The identified signs and symptoms which are stored in the knowledge base are searched to identify the disease.
3. Getting the weighing factors (wf) (the associated degree of intensity) $wf = 1$, 2, 3, 4 where 1 = low, 2 = average, 3 = high and 4 = very high.

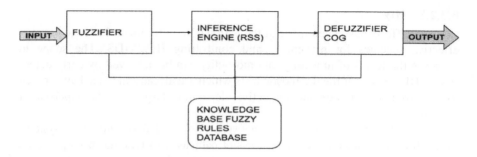

FIGURE 8.5 A representation of a fuzzy expert system

4. Applying fuzzy rules to map fuzzy inputs into their corresponding weighting factors.
5. Determining the rule base.
6. Determining the firing strength of the rules.
7. Calculating the degree of truth R, of each rule with the help of evaluation of a minimum value that is greater or equal to zero.
8. Computing the severity of the disease.
9. Obtaining the resultant value.

8.3.2.7 Diabetes

Diabetes, also known as diabetes mellitus, causes high blood sugar. With diabetes, two cases are likely to arise: the body not producing appropriate levels of insulin or ineffective use of the insulin it produces. Insulin is a highly vital hormone for the proper functioning of the body. Several factors including genetics, lifestyle or hormonal diseases can affect insulin levels in a body. Thus, fuzzy systems are an efficient medium to monitor and diagnose diabetes. The development of the diagnosis mechanism of diabetes is done using FES [25]. Fuzzy expert system modelling is pursued using the following steps:

1. Selecting the input and output variables.
2. Determining the fuzzy number.
3. Fuzzification interface.
4. Choosing the appropriate membership function and fuzzy operators.
5. Designing the correlation fuzzy for determination mechanism with determination logic and knowledge base.
6. Defuzzification interface.

The input variables are plasma glucose concentration in 2-hours OGTT(Glucose), 2-hour Serum Insulin(INS), Body Mass Index(BMI), Diabetes Pedigree Function (DPF), Age(Age) and the output variable is Diabetes Mellitus(DM). The inputs are transformed into the fuzzy set. The conversion of each input to its fuzzy equivalent is carried out. Additionally, an interface is used for the tuning and validation of

Fuzzy Systems in Medicine and Healthcare 153

the parameters of the built fuzzy numbers. The membership values and degrees, obtained by the fuzzification of the crisp input values, are formulated in correlation with the fuzzy determination mechanism. Here, the defuzzification unit receives the output values obtained by using the correlation fuzzy determination mechanism that can take fuzzy value inputs. The mapping from the input variable to the output variable is implemented by associating the fuzzy determination mechanism with logic and knowledge base. The rule base consists of nine if-then rules. The fuzzy operator is applied in case the antecedent of a given rule has more than one part. The evaluation of the antecedent of the rule takes place. The defuzzification process helps to transform the result into a crisp value for DM output.

8.3.2.8 Asthma

Asthma, a chronic inflammatory lung disorder, has been known to affect 1.4–27.1% of the population worldwide. The underdiagnosis of asthma, especially in developing countries, has been proven in various studies. The diagnosis of asthma at early stages has been successful with the development of fuzzy expert systems. The knowledge base containing declarative and procedural knowledge in addition to literature knowledge (textbooks and papers) is employed during the diagnosis process. Knowledge representation of this system is characterised in two forms called Type A and Type B. Type A, which is made up of six modules, incorporates symptoms, allergic rhinitis, genetic factors, symptom hyper-responsiveness, medical factors and environmental factors. Type B, which is made up of eight modules, incorporates symptoms, allergic rhinitis, genetic factors, response to short-term drug use, bronchodilator tests, challenge tests, PEF tests and exhaled nitric oxide. The final result provides an assessment of the possibility of asthma for the patient [26].

8.3.3 Monitoring Patient's Condition during Heart Surgery

The tracing of vital functions of the patient during coronary bypass surgery (CBS) has been enabled with the designing of the FES. The designed fuzzy expert system ensures the presentation and monitoring of the vital functions (blood pressure, haemoglobin, pulse, and beta-blocker). The interpretation of these functions to indicate the condition of the patient is carried out. The output of the system is presented in audio and visual forms. The data collected from the supervised fuzzy expert system and inferences known from earlier literature or operator's experience are compared. Then the returned output frequency value is interpreted by the system. The integration of a lamb into fuzzy expert software as a visual indicator is also done. Colours of the lambs such as blue, green, yellow, orange and red represent the condition levels.

8.4 ROLE OF FUZZY SYSTEMS IN MEDICINE

Since the 1970s, a great deal of research and literature has been done to explore the potential of fuzzy logic in medicine. Accordingly, medical data analysis shifted steadily towards computer-based applications. Albin and Sanchez produced some of the earliest works in the integration of fuzzy set theory in medical knowledge-based systems, which provided a foundation for future research and experimentation.

It should be noted that, historically, a lot of seemingly vague medical terms were considered precise or simply not a fuzzy concept. For a general example, high fever might correspond to a 100°–104° Fahrenheit temperature range or blood sugar may be above normal if it exceeds 140 mg/dL. Thus, a lot of discussions and investigatory work had to be done to incorporate existing medical linguistics with rapidly growing soft computation techniques such as Fuzzy logic. After the works of Sadegh-Zadeh in establishing the role of fuzzy logic as the ideal type for healthcare and medicine [27] and R. Seizing, Tabacchi et al.'s publication of 'Fuzziness and Medicine: Philosophical Reflections and Application Systems in Health Care' [28], fuzzy logic steadily became an accepted part of medical analytics and knowledge-based systems. Today, fuzzy logic coupled with other soft computing techniques is extensively used in operation theatres to mitigate risk in surgeries, monitoring devices and many more critical areas of health and medicine.

A physician or practitioner assesses risk factors for patients based on subjective and ambiguous circumstances and/or additional experience-based understanding of the situation. In this era of innovation, the majority of the practitioners work with computer-based information systems in the diagnosing process. These systems are either for them to record the data or support their decision-making process. In the given methods, intensive volume and various types of data can exist. Thus, special cases and unique situations require specific attention in the process. In this section, several methodologies are discussed which are incorporated with fuzzy logic systems in analysis and decision-making processes.

8.4.1 ANALYSIS OF SIGNS OR SYMPTOMS

Let's reconsider our previous example of a diagnosis of blood sugar levels. It should be noted that a blood sugar level up to 130 mg/dL is considered normal. But there are many other factors which should affect the diagnosis. A simple two-valued logic system will declare 'HIGH' or 'LOW' based only on the measured blood sugar level, which might ignore other substantial parameters such as patient's diabetic history, diet, or time since last food intake. Another case that we can study is a diagnosis of breast cancer. Its signs and symptoms include lump formation in the breast or near underarm area, change of texture or size of the breast, swelling, and so on. Also, a high risk of breast cancer exists if a relative or a member in the previous generation was also diagnosed. It is advised that early detection of the indication of cancer growth is necessary for successful treatment and recovery.

Throughout this chapter, we have seen the importance of high accuracy systems and the need to integrate multiple variables, both known and unknown, in medical decision-making systems. Our efforts with any diagnosis or prognosis apparatus should be of prevention of proliferation of ailment or risk minimisation. Human operators base their analysis on the basis of their experience, existing methodology or inherent logical process. A major responsibility of the people in this profession is to study inferences and provide an understanding of a situation on the basis of adequate reasoning. However, the possibility of human error cannot simply be eliminated from the equation. Thus the development of highly accurate and reasonable decision support systems for analysis or diagnosis is pivotal in the medical industry.

Fuzzy Systems in Medicine and Healthcare

In this section, we will study the role of soft computing techniques in various existing methodologies used extensively in day-to-day healthcare.

8.4.2 MEDICAL IMAGE PROCESSING

Fuzzy set theory is extensively used in the mathematical framework of various steps of image processing. It provides a robust way of representing structural information and analysis of both numerical and symbolic data. Real-world image processing applications are characterised by their vagueness and uncertainty. Particularly in the medical domain, MRI scans, ultrasound images or CT scans may yield inconsistent or imprecise results either due to system limitations or anatomical ontology. Several important steps in image processing such as contour detection, indexing, retrieval and cleaning of image data are optimised with the applications of fuzzy techniques. This is achieved by representing image information within different levels of fuzzy sets. Image segmentation, an essential application of image processing, can be highly improvised using the fuzzy C-means clustering method. This process is also effectively applicable to the detection of lesions in brain scans. Another important application of incorporating fuzzy techniques with medical image processing is a fusion of images or scans from different sources, for example fusing individual scans of soft and dense tissues to aid diagnosis process. Different scans such as MRI or tomography highlight different regions in a scan. Thus, to obtain higher accuracy or detailed highlights, the fuzzy-based image fusion process is used [29]. The process has proven superior in comparison to traditional methods, as fuzzy techniques provide the capability to assess distorted or even missing data, provided an adequate knowledge base is utilised. In the absence of expert opinion, medical data or historical knowledge can be used to integrate the ANFIS technique with image processing.

8.4.3 ADMINISTRATION OF ANAESTHESIA

Fuzzy logic provides the capability to monitor anaesthesia control. Fuzzy systems dedicated to anaesthesia control have to take into consideration multiple parameters to adopt with the patient's condition. Different humans respond differently to anaesthetics or similar stimulus. Thus, the control systems used in anaesthetic control should be fast, accurate and reliable. More importantly, they should be adaptive to human nature, of both the operator and the patient. Anaesthetic decision support systems are generally of multi-input single-output (MISO) nature. Various factors such as muscle relaxation, amount of anaesthetic drug, heart rate or blood pressure are calculated to determine the further course of drug delivery. Along with the effectivity of the system, the experience of the anaesthetist is also a contributing factor to a successful administration and outcome of the medical procedure. There can be situations where the operator himself could realise the effects of the anaesthetic drug if inhaled indirectly or accidentally, which might further impact drug delivery to the patient. It should be noted that all the factors mentioned are dynamic in nature—that is, they change over the duration of anaesthesia administration. In this case, regular feedback input is essential for an effective controller mechanism. In the control loop, the anaesthetic is the controlling block while the patient is the feedback component.

Intelligent systems developed with fuzzy systems are highly regarded for their reliable performances both as primary controllers for the administration of anaesthetic drugs or as supporting systems for a human operator [30].

8.4.4 Determination of Drug Dose

The body's reaction to a drug and interaction of different drugs in case of multi-drug prescriptions is critical to the treatment and health of a patient. Though standardised systems and methodology exist for determining a drug and its amount to be prescribed, the patient's physiology and medical history are also pertinent factors which might often not be weighed correctly by a human administrator [31]. Thus, intelligent expert systems are used to minimise medication errors by incorporating human instincts with expert knowledge and algorithmic computations. Intelligent prescription systems are used to formulate patient-specific drug routines. It is very common for a doctor to prescribe different drugs for the very same medical condition. This decision is influenced by many factors such as the severity of the disease, previous medications and any notable reaction to a similar drug. Fuzzy-based intelligent systems can be utilised to provide further insights such as the possibility of any adverse drug reactions based on the medical data or expert knowledge used in developing the system [32]. These systems are helpful in scenarios where there is uncertainty or lack of confidence in the decision of the doctor or practitioner.

8.4.5 Improving Service Quality

Till now, we have observed how different fuzzy techniques have been modelled or utilised in amalgamation with other soft computing techniques to directly contribute in medical procedures like diagnosis or surgery, or as decision support systems to aid human operators. Now this section highlights the use of fuzzy logic-based systems or tools in the domain of clinical services or health management systems. In recent years, fuzzy-based data management systems and service quality evaluation and management systems are gradually becoming popular and more widely accepted in hospitals and medical clinics (see Figure 8.6).

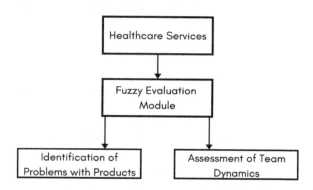

FIGURE 8.6 Framework for implementing fuzzy logic in service evaluation.

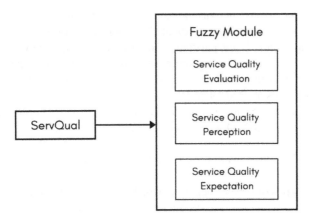

FIGURE 8.7 Representation of parameters in a fuzzy evaluation module.

This growth can be attributed to the fact that fuzzy methods allow effective computation of variables such as patient satisfaction, hospital environment, food quality, security or medical staff's responsivity, which seems highly improbable with any other methodology. Strategic efforts in improvising a particular department or developing effectual resource allocation plans are possible with the correct assessment of service feedback data. Fuzzy SERVQUAL is a popular tool used by managers [33] to study the performance of their teams and explore possibilities in capacity building by minimising redundancy or gaps in service quality (see Figure 8.7).

8.4.6 Service Quality Evaluation in Yoncalı Physiotherapy and Rehabilitation Hospital—Case Study

Yoncalı Physiotherapy and Rehabilitation Hospital (YPRH) was established in 1991 as a public hospital in Kütahya, Turkey. The hospital is known to offer various services such as inpatient, outpatient, polyclinic and radiology with a capacity of accommodating over 200 patients. The fuzzy SERVQUAL method was put to use in order to determine and evaluate the patient's satisfaction status. Five steps are followed as a part of the solution process: the problem statement, designing a survey, developing a method of data acquisition, examination and administration of results. The inpatient services are defined in the first step of the process for evaluating patients' satisfaction status. The questionnaire consisted of two parts. One of the parts included demographic variables such as gender, age group, level of education and monthly remuneration. The other SERVQUAL section of the questionnaire included 22 items within 5 dimensions. Linguistic variables (which consisted of strongly disagree, disagree, neither agree nor disagree, agree, strongly agree) were used and converted into triangular fuzzy numbers for the purpose of analysis. The regulation of admissions is done for sample time periods with respect to the physiotherapy treatments. The questionnaires were applied to 262 patients, and interviews were held for the patients.

158 Soft Computing Applications and Techniques in Healthcare

Cronbach's alpha was put to use for the testing of the uniformity and reliability of the obtained data. The outcomes indicated an overall reliability score of 0.81 for all dimensions and over 0.71 for particular dimensions. An alpha of 0.70 is considered acceptable, and this helped in the confirmation of the reliability of the obtained data. The YPRH administration wanted to keep expectation scores as high as possible in order to establish a standard for improving hospitals.

In conclusion, the service quality gap can be calculated as follows:

G1 = (3.47, 4.26, 4.83)–(4.5, 5, 5) = (3.47 – 5, 4.26 – 5, 4.83 – 4.5) = (–1.53, –0.74, 0.33).

Some of the key findings of the case included the following:

- All the gap scores were negative, which was indicative of the general dissatisfaction with regard to the services being provided.
- Empathy scores were the highest, followed by assurance, reliability, responsiveness and tangibles.
- The cleanliness of staff's appearance had the highest score followed by hospitals' modern equipment availability with regard to the perception of patients towards tangible dimensions.
- The timely manner in treatments had the highest score, followed by the willingness factor in helping patients
- The priority of patients' benefits and trust in the secure nature of personal information and diagnosis earned the highest score in the assurance dimension.

Finally, all results with their interpretations were presented to the hospital administration, where they were verified by the hospital administration.

8.5 CONCLUSION

Recent trends show that the utilisation of fuzzy logic techniques in the field of healthcare has helped to improve the decision making of health professionals, thus improving the overall condition of the healthcare services. This chapter elaborated on the various fields where soft computing techniques have gained significant recognition. The application of fuzzy logic in the diagnosis of disease not only helps to produce timely solutions to some of many critical complications but also helps physicians to better understand the condition of the patient. The ultimate goal of every fuzzy technique employed to the various healthcare sectors is to improve the accuracy of the outcomes.

Fuzzy logic finds its applications in the easy and extensive study of medical databases using the rough set theory and fuzzy logic. The use of such an approach not only removes the ambiguity in the medical databases but also provides a scope of improved interpretation of data, leading to a better study of the various trends related to the changing nature of the disease in recent times. Information retrieval has also become easier with the employment of keyphrase and keyword extraction techniques to improve upon the vagueness of the various existing retrieval systems.

Fuzzy Systems in Medicine and Healthcare

The widespread use of fuzzy logic in the diagnosis and treatment of various diseases has helped to reduce the mortality rates and provide an enhanced treatment experience to patients worldwide. The numerous cases of various types of cancer plaguing the health of billions have gained significant focus in terms of research. Many research studies aim toward better detection and treatment of such deadly diseases. Lung cancer, being one of the widespread types of cancer, can be detected more efficiently with the help of a lung cancer diagnosis system. Because the nature of every type of cancer results from the abnormal multiplication of cells, the fuzzy techniques employed to each cancer type is also different. Breast cancer has been known to be the most common cause of fatality among women. The combined use of fuzzy rules and thermal imaging helps medical practitioners to better detect the disease and provide a timely solution for the same. The level and stage of breast cancer using fuzzy techniques has helped to improve upon the treatment services in many medical institutions. Diseases such as periodontal dental disease can also be diagnosed with the aid of Mamdani approach as an output mechanism and Mamdani max-min rule to formulate the validity degree for each if-then rule. Anaemia, a widespread disease among the children belonging to the underprivileged sections of the society, can also be detected to save thousands of lives which are at risk of being prone to this disease. The utilisation of triangular membership functions has facilitated the easy gathering of information, thereby leading to a better case building of the patient. The use of a neuro-fuzzy system to detect HIV levels and employing of the if-then rules to further improve the efficiency of the results has long been in use.

The algorithm described in this chapter for the detection and treatment of malaria has been studied by many researchers. It not only helps in the timely diagnosis of the disease but also indicates the level of the disease, thereby helping practitioners to improve upon their decision-making process and restoring the health of the patient in the best way possible. Several fields such as diabetes, asthma and monitoring the medical condition of the patient during heart surgery have also experienced an improvement in their process with the help of developed rules and fuzzy expert systems.

The existing use of fuzzy logic has explored various areas where its employees can increase the efficiency of the outcomes and provide clarity to the information gathered. However, it can still be employed in various areas to optimise their treatment methods. There are many diseases which have not been studied with regard to fuzzy logic and can be worked upon in order to better the delivery of services and treatments. This chapter elaborately discussed the future scope of fuzzy systems and related soft computing techniques to aid medical institutions to achieve better results in a timely and patient-friendly manner.

REFERENCES

1. Ross, T. J. Fuzzy Logic with Engineering Applications. John Wiley & Sons Ltd, The Atrium, Southern Gate, Chichester, 2004.
2. Běhounek, L., and Cintula, P. (2006). 'From fuzzy logic to fuzzy mathematics: a methodological manifesto'. Fuzzy Sets and Systems. 157(5): 642–646.
3. Dvǒřak, A., and Novak, V. Fuzzy logic as a methodology for the treatment of vagueness. In L. Běhounek and M. Bílkova (eds.), The Logica Yearbook 2004. Filosofia, Prague, 2005, pp. 141–151.

4. Konar, A. Computational intelligence: principles, techniques and applications, Springer, Berlin Heidelberg New York, 2005.
5. Bezdek, J. C. Fuzzy mathematics in pattern classification. PhD Thesis, Applied Mathematics Centre. Cornell University, Ithaca, 1973.
6. Dubois, D., Prade, H. Fuzzy Sets and Systems: Theory and Applications. Academic Press, New York, 1980.
7. Mahfouf, M., Abbod, M. F., and Linkens, D. A. (2001). 'A survey of fuzzy logic monitoring and control utilization in medicine'. Artif. Intell. Med. 21: 27–42.
8. Seising, R. (2004). 'A history of medical diagnosis using fuzzy relations'. Fuzziness in Finland'04, 1–5.
9. Nguyen, H. P. and Vladik, K. (2001). 'Fuzzy logic and its applications in medicine'. Int. J. Med. Inform. 62: 2–3.
10. Aiken, L. H., Sermeus W., Van den Heede, K., et al. (2012). 'Patient safety, satisfaction, and quality of hospital care: cross-sectional surveys of nurses and patients in 12 countries in Europe and the United States'. BMJ. 344: e1717
11. Collins, S. R., Radley, D. C., Schoen, C., and Beutel, S. National trends in the cost of employer health insurance coverage, 2003–2013, Commonwealth Fund, December 2014.
12. Prasath, V., Lakshmi, N., Nathiya, M., Bharathan, N., and Neetha, N. P. (2013). 'A survey on the applications of fuzzy logic in medical diagnosis'. Int. J. Sci. Eng. Res. 4: 1199–1203.
13. Papageorgiou, E. I. (2012). 'Learning algorithms for fuzzy cognitive maps - a review study'. IEEE Trans. Syst. Man Cybern. C: Appl Rev. 42: 150–163.
14. De, S. K., Biswas, R., and Roy, A. R. (2001). 'An application of intuitionistic fuzzy sets in medical diagnosis'. Fuzzy Sets and Systems. 117(2): 209–213.
15. Melin, P., Castillo, O., Kacprzyk, J. Design of Intelligent Systems Based on Fuzzy Logic, Neural Networks and Nature-Inspired Optimization, Vol. 601. Springer, Switzerland, 2015.
16. Lauraitis, A., Maskeliūnas, R., and Damaševičius, R. (2018). 'ANN and fuzzy logic based model to evaluate Huntington disease symptoms'. J. Healthcare Eng. 2018: 10. Article ID 4581272. https://doi.org/10.1155/2018/4581272.
17. Yadollahpour, A., Nourozi, J., Mirbagheri, S. A., Simancas-Acevedo, E., and Trejo-Macotela, F. R. (2018). 'Designing and implementing an ANFIS based medical decision support system to predict chronic kidney disease progression'. Front Physiol. 9: 1753. doi:10.3389/fphys.2018.01753.
18. Dev, U., Sultana, A., Talukder, S., and Mitra, N. K. (2011). 'A fuzzy logic approach to decision support in medicine'. Bangladesh J. Sci. Ind. Res. 46(1): 41–46.
19. Harvey, N. R., and Marshall, S. (1995). 'The design of different classes of the morphological filter using genetic algorithms', IEEE fifth international conference on image processing and its applications.
20. Aguilera, P. A., Fernández, A., Reche, F., and Rumi, R. (2010). 'Hybrid Bayesian network classifiers: application to species distribution models'. Environ. Model. Software. 25(12): 1630–1639.
21. Hanea, A., Kurowicka, D., Cooke, K., and Ababei, D. (2010). 'Mining and visualising ordinal data with non-parametric continuous BBNs'. Comput. Stat. Data Anal. 54(3): 668–687.
22. Specht, D. F. Probabilistic neural networks for classification, mapping, or associative memory. Proceedings of IEEE International Conference on Neural Networks, Vol.1. IEEE Press, New York, pp. 525–532, June 1988.
23. Usuki, H., Onoda, Y., Kawasaki, S., Misumi, T., Murakami, M., Komatsubara, S., et al. (1999). 'Relationship between thermographic observations of breast tumours and the DNA indices obtained by flow cytometry'. Biomed. Thermol. 10(4): 282–285.

Fuzzy Systems in Medicine and Healthcare

24. Tunali, M., Clinical Examination, Diagnosis, Prognosis and Treatment Planning in Periodontology, http://nnno.facebook.com/topic.php?uid=18402237696&topic=3998 (05.02.2010).
25. Lee, Chang-Shing (Feb. 2011). 'A fuzzy expert system for diabetes decision support application'. IEEE Trans. Syst. Man Cybern. B: Cybernetics. 41(1): 139–153.
26. Bugiani, M., Carosso, A., Migliore, E., Piccioni, P., Corsico, A., and Olivieri, M. (2005). 'Allergic rhinitis and asthma comorbidity in a survey of young adults in Italy'. Allergy. 60:165–170.
27. Sadegh-zadeh, K. (2000). 'Fuzzy Health'. Illness Disease J. Med. Philos. 25: 605–638.
28. Sadegh-Zadeh, K. (2001). Fuzziness and Medicine: Philosophical Reflections and Application Systems in Health Care, DOI: 10.1007/978-3-642-36527-0, R. Seising and M. E. Tabacchi (Eds.), Series ISSN: 14349922, Springer-Verlag Berlin Heidelberg.
29. Seising, R. (2006). 'From vagueness in medical thought to the foundations of fuzzy reasoning in medical diagnosis'. Artif. Intell. Med. 38: 237–257.
30. Neukirchen, M., and Kienbaum, P. (2008). 'Sympathetic nervous system: evaluation and importance for clinical general anaesthesia'. Anesthesiology. 109: 1113–1131.
31. Boyd, C. M., Darer, J., Boult, C., Fried, L. P., Boult, L., and Wu, A. W. (2005). 'Clinical practice guidelines and quality of care for older patients with multiple comorbid diseases: implications for pay for performance'. JAMA. 294(6):716–724.
32. Gawedal, Adam E., Jacobs, Alfred A., Brierl, Michael E. Fuzzy Rule-based Approach to Automatic Drug Dosing in Renal Failure, The IEEE International Conference on Fuzzy Systems, 2003.
33. Lupo, T. (2013). 'A fuzzy ServQual based method for reliable measurements of education quality in Italian higher education area'. Expert Syst. Appl. 40(17): 7096–7110.

9 Appliance of Machine Learning Algorithms in Prudent Clinical Decision-Making Systems in the Healthcare Industry

T. Venkat Narayana Rao and G. Akhila
Sreenidhi Institute of Science and Technology
Hyderabad, Telangana, India

CONTENTS

9.1 Introduction 164
    9.1.1 Artificial Intelligence.. 164
    9.1.2 Healthcare Industry 165
9.2 Computers and Healthcare Supervision in New Era 165
    9.2.1 Medication 166
    9.2.2 Patient Diagnosis 167
    9.2.3 Telemedicine... 167
    9.2.4 Surgical Procedures.. 167
    9.2.5 Communication and Sharing Information.................................... 167
    9.2.6 Medical Imaging... 168
9.3 Importance and Emergence of Machine Learning.................................... 168
    9.3.1 Supervised Learning.. 168
    9.3.2 Unsupervised Learning 170
    9.3.3 Reinforcement Learning.. 170
    9.3.4 How Machine Learning Has Changed Health Aspects.................. 170
9.4 Automatic Drug Discovery.. 171
    9.4.1 Stages Involved in Drug Discovery 171
9.5 Feature Selection in Medicine 171
    9.5.1 Attribute Ranker 172
    9.5.2 Principal Component Analysis 172
    9.5.3 Factor Analysis 173
    9.5.4 Median Imputation 173

164   Soft Computing Applications and Techniques in Healthcare

9.6	Regression Analysis in Medicine	173
	9.6.1 Application of Regression in Medicine	174
9.7	Querying in Medicine	174
9.8	Density Estimation in Medicine	175
9.9	Dimension Reduction in Medicine	176
	9.9.1 Feature Evaluator and Feature Ranker Algorithm	177
	9.9.2 Linear DR Algorithms	177
	9.9.3 Nonlinear DR Algorithms	177
	9.9.4 Clustering Algorithms	177
9.10	Testing and Matching in Medicine	178
	9.10.1 Diagnostic	178
	9.10.2 Screening	178
	9.10.3 Monitoring	179
	9.10.4 Precision or Accuracy Aspects	179
9.11	Classification and Clustering in Medicine	179
	9.11.1 Clustering	179
	9.11.1.1 Hierarchical Clustering	179
	9.11.1.2 Partition Methods	180
	9.11.1.3 Density-Based Clustering	181
	9.11.1.4 K-Nearest Neighbour	181
	9.11.2 Classification	182
	9.11.2.1 Texture Classification	182
	9.11.2.2 Neural Networks	182
	9.11.2.3 Data Mining	183
9.12	Conclusion	183
References		183

9.1 INTRODUCTION

The world is now full of inventions and discoveries. Artificial intelligence and machine learning are playing a vital role in this modern era and has laid a path to these new discoveries and inventions in every field. As a part of it, many new advancements and research are going on in the healthcare industry. The healthcare industry is one of the major and fastest growing sectors in the world. It offers services and goods to treat people with precautionary, palliative, rehabilitative and curative care. The healthcare industry often uses various machine learning algorithms and artificial intelligence to store and analyse large volumes of data for better decision making.

9.1.1 ARTIFICIAL INTELLIGENCE

Artificial intelligence, also called machine intelligence, is one of the diverging areas in today's world. It aims at extending and supplementing the efficiency and capacity of humans in tasks such as governing society and imitating nature through machines, the final goal of AI is to make humans and machines coexist harmoniously in the world. It is frequently employed to depict machines that imitate cognitive traits of

Machine Learning Algorithms in Clinical Decision-Making Systems 165

the human brain such as learning and problem solving. These traits are completely distinct to natural intelligence of humans.

AI can be categorised into three diverse types such as human-inspired, analytical and humanised:

1. Human-inspired: This is the combination of emotional and cognitive intelligence. Here, emotional intellect represents perceptive human feelings. Both cognitive and emotional elements are considered in order to make decisions.
2. Analytical AI: This consists of uniqueness that is associated with cognitive intelligence. Cognitive intelligence is a way of making future decisions based on past experiences, which helps in producing a cognitive illustration of the world.
3. Humanised AI: This represents all types of characteristics such as cognitive, emotional and social intelligence.

Artificial intelligence was founded as a scholarly discipline in 1956. The conventional problems of AI research comprise knowledge representation, reasoning, planning, learning, perception, natural language processing and the capability to shift and control objects. General intelligence is among the domain's long-term objective. The approaches include statistical process, computational intelligence and conventional symbolic AI. Numerous tools are utilised in AI, including alternatives of search and artificial neural networks, mathematical optimisation and mechanisms based on statistics, economics and probability. The AI domain deals with computer-science, information, psychology, linguistics, engineering, mathematics, philosophy and various other areas.

9.1.2 Healthcare Industry

The healthcare industry is an integration and collection of various economic sectors which help in providing goods and services to patients. It contributes more than 10% of gross domestic product among many developed countries in the world and involves production and commercialisation of services and goods to automate, maintain and establish health. The current healthcare business is divided into several sectors, and it depends on the multidisciplinary groups of skilled professionals and paraprofessionals to meet the health needs of humans. It is currently in the process of a complete global overhaul and transformation. Figure 9.1 shows the development of medical practices in healthcare using AI.

9.2 COMPUTERS AND HEALTHCARE SUPERVISION IN NEW ERA

In the area of medicine, computers permit more rapid transfer of information between doctors and patients. Computers are an exceptional tool for storage of patient-associated data. Large hospitals utilise computer systems to keep patient records, and computer systems can keep track of prescriptions and billing data. Computers can store information concerning the medication prescribed to a patient and those which cannot be prescribed to the patient. Hospital-bed alert systems, emergency alarms, X-ray machines and numerous other medical equipment are based on computer system logic only. The software is used to diagnosis the disease. A few multifaceted surgeries can be done with the help of computers. In addition, several modern methods of imaging and scanning are mainly based on computer tools.

FIGURE 9.1 AI in healthcare.

There are six common areas in which computers play a significant role in the healthcare industry: medication, patient diagnosis, telemedicine, surgical procedures, communication and sharing information and medical imaging. The uses are explained in the following sections.

9.2.1 Medication

Nowadays hospitals are using computers to store patient records. Doctors review the health records in order to prescribe medication to the patients. In addition, they use software programs to manage these records. This helps in preventing human error.

Computers also help in retrieving, updating and securing patient records. Figure 9.2 shows how computers are used in the area of medication.

FIGURE 9.2 Use of computer in medication.

9.2.2 Patient Diagnosis

If multiple doctors are treating the same patient, it is necessary for them to know about the patient. As the patient's records are already stored in computers, doctors use this stored information for the proper diagnosis of the patient. This makes accessing the details of patient effortless and hassle-free.

9.2.3 Telemedicine

Many people in rural areas have very poor medical facilities and cannot afford proper healthcare. This gap between those with low income and doctors can be filled using technologies. Telemedicine enables a patient to directly interact with the practitioners via video call or phone. Healthcare professionals can then advise whether or not the person needs medical help. In addition, it can be invaluable for people with disabilities and the elderly, who often have mobility issues. Telemedicine also records and stores the information electronically. Figure 9.3 shows an example of telemedicine.

9.2.4 Surgical Procedures

Today, computer-assisted surgery is becoming increasingly popular among surgeons performing many advanced surgical procedures. It is the combination of computer intelligence and human intelligence which helps assist surgeons during many complex operations. Computer-assisted surgery systems helps create creating a model through the analysis of patient data; doctors use this model to generate the surgical procedures. Figure 9.4 is such surgical procedure.

9.2.5 Communication and Sharing Information

Computer technology has become one of the most common tools for communication. In healthcare, it enables patients to interact with doctors. In addition, many medical periodicals, diagnosis manuscripts, significant medical documents and research and reference books can be stored in digital form to make them available to users.

FIGURE 9.3 Telemedicine.

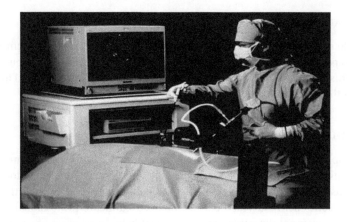

FIGURE 9.4 Surgical procedure.

9.2.6 Medical Imaging

Medical imaging is a technique where a machine can create visual representations of interior body parts, which helps in analysis and diagnosis. It can represent the function of some tissues and organs; it also represents internal structures that are hidden in the skin and bones.

9.3 IMPORTANCE AND EMERGENCE OF MACHINE LEARNING

Machine learning (ML), a domain of statistical models and algorithms, helps computer perform many tasks using patterns and inferences without any explicit instructions. It is a subset of artificial intelligence, as shown in Figure 9.5.

In machine learning, decisions are made with the help of algorithms that aid in building a mathematical model for the given sample data, called training data.

Following are three types of machine-based learning:

- Supervised learning
- Unsupervised learning
- Reinforcement learning

Figure 9.6 shows various learning techniques through machine learning.

9.3.1 Supervised Learning

In supervised learning, each example is a pair of one input value and one output value. This data is analysed using algorithms and helps in creating an inferred function, which further helps in the mapping of new examples.

Machine Learning Algorithms in Clinical Decision-Making Systems 169

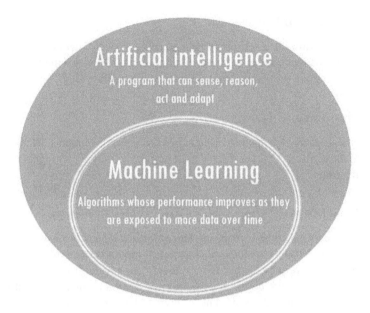

FIGURE 9.5 Relation between artificial intelligence and machine learning.

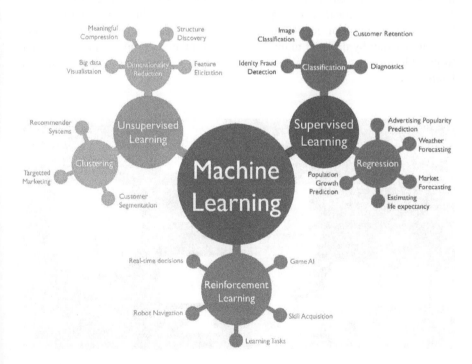

FIGURE 9.6 Machine learning and its branches.

9.3.2 Unsupervised Learning

Unsupervised learning is a self-organising learning system where one can find the unknown patterns in a dataset without existing labels. The major methods in unsupervised learning are principal component analysis and clustering.

9.3.3 Reinforcement Learning

Reinforcement learning helps in taking actions where punishments are avoided or reward are maximised. This learning is most popularly used in gaming and rea-time decision making.

9.3.4 How Machine Learning Has Changed Health Aspects

There are many areas where machine learning techniques are used, but one in particular where ML has a widespread societal impact is in healthcare industry. One of the best examples of this is increase in the production of smart bands, which are devices which continuously monitor the user's health condition like heartbeat, sleep time and so on. These devices have bridged the gap between rising healthcare costs and low-income people and also contribute in building a better relationship between doctors and patients. Many new techniques for the detection and prediction of disease in the early stages are being introduced. The best example for this is the prediction of cancer, where machine learning has reduced the rate of deaths due to cancer. The healthcare industry is growing rapidly day by day and many new advancements are taking place in order to provide proper care for the patients. Figure 9.7 shows the impact of ML on healthcare.

FIGURE 9.7 Machine learning in healthcare.

Machine Learning Algorithms in Clinical Decision-Making Systems

9.4 AUTOMATIC DRUG DISCOVERY

The process of drug discovery has had a huge impact on the pharmaceutical industry and in life science. It offers great benefits ranging from web services to semantic web technology. Over the past few decades, CADD (computer-aided drug discovery) has evolved enormously in the field of mature science. The main aim of drug discovery is to use computational approaches to select the candidate molecules and for further access of their suitability as a drug target, which also includes detrimental effects. There are many fields involved, such as clinical and biological domain, chemical domain, chemical structure processes, information from different sources, databases, microanalysis, screening, combinatorial chemistry and so on. Because the drug discovery is interdisciplinary in nature, it integrates all these domains.

Most of the proteins in the human body are enzymes. These enzymes speed up the rate of chemical reaction without being irreversibly changed themselves. Others involve controlling highly specific cell adaption and reactions. They work with a principle called lock and key, where one or more smaller molecules (substrates, ligands) fits comfortably into a hole binding pocket, which is an active site of a protein, thereby facilitating a subtle structure change. This in turn leads to a chain reaction.

The binding of drug molecules is made to be structurally similar native ligands (candidate keys). These are fake duplicate keys, close enough to fit the hole but not of the right shape to turn the block. Most of the current drugs are 'small molecules'. Apart from that, other classes of medicines exist and are being developed, such as biologic drugs or therapeutic antibodies.

9.4.1 STAGES INVOLVED IN DRUG DISCOVERY

- Target selection and validation
- Hit discovery
- Hit to lead
- Lead optimisation
- Preclinical development

9.5 FEATURE SELECTION IN MEDICINE

Medical databases consist of large sets of data that are related to disease makers, which help in the prediction of disease. But at times some of the disease makers are not helpful and may also have a negative impact on the results. So, generally, in order to advance the precision of the model's output, it assumed to remove the extraneous features in the data and identify only the related features in the data, which is a very difficult task for a human. We can overcome this problem by implementing feature selection techniques. Feature selection is one of the central theories of machine learning, and it has a major impact on the performance of the model. It also helps in making effective medical decision making [1, 4].

Feature selection is the first preprocessing technique in designing a model, which helps in identifying essential attributes and enables a major function in the course of classification. It is a process where one can select the features manually or automatically that are required for prediction. Having relevant features in the data increases the accuracy of the models, whereas irrelevant features in data decreases its accuracy and makes the model learn based on its irrelevant features. The training set in machine learning models has a great influence on the performance [2].

Following are the merits of performing feature selection prior to modelling the data:

- Reduces overfitting: Decision making becomes easy, as there will be less redundant data.
- Improves accuracy: Accuracy increases because misleading data has been reduced.
- Diminishes training time: Fewer data points reduces algorithm's complexity and algorithms train swiftly.

So, how can we select features? There are many techniques by which features can be selected. Following are some of these:

1. Attribute ranker
2. Principle component analysis
3. Factor analysis
4. Median imputation

9.5.1 ATTRIBUTE RANKER

The attribute ranker algorithm evaluates a subset containing one or more features in it and gives a goodness score using a subset evaluator [2]. Similarly, the attribute evaluator evaluates every individual attribute. When search algorithm and sunset evaluator are paired, it explores the subsets in the space and returns the best evaluated subset as a result. Later, a ranked list is produced by pairing the special search called ranker and the attribute evaluator. From that ranked list, the top M ranked attributes are selected for further processing.

9.5.2 PRINCIPAL COMPONENT ANALYSIS

Principal component analysis (PCA) is a useful linear dimension reduction technique. It has a nature of a second-order method which is based on the covariance matrix of the variables. This method basically consists of the following four steps [2]:

1. Normalisation of input data is done to ensure that there is no domination of attributes with a large domain on attributes with a small domain.
2. The PCA method helps in computing K orthogonal vectors which are used as standard components.

Machine Learning Algorithms in Clinical Decision-Making Systems 173

3. These principal components are sorted in a descending pattern of strengths or significance. This step helps provide more important information about the variance.
4. After sorting, the data is reduced by removing the weak significant components.

9.5.3 FACTOR ANALYSIS

Factor analysis is a linear method, which relies on summaries of second-order data just like the PCA method. This method assumes that there are some unknown variables on which the measured variables depend. The aim of this method is to find the relations between those measured and unknown variables by which one can reduce the dimensions of the datasets.

9.5.4 MEDIAN IMPUTATION

The median imputation method is best when the features are skewed. The presence of outliers in the data affects the mean. Hence, the use of median instead of mean assures strength. In this method, the missing data of a specified trait is substituted using the median of the known values of the attributes that are in the class of the missing data.

9.6 REGRESSION ANALYSIS IN MEDICINE

Regression analysis is a part of supervised learning in machine learning and consists of a set of methods which allows a user to predict a continuous variable (y) that is dependent on the value of one or multiple predictor variables (x). Regression analysis aims at constructing a mathematical equation in which y is defined as a function of x.

In medicine, it is often observed that the effect of statistics is applied on design part of the study. The traditional statistical methods are not useful for the proper study design; hence, it has become necessary to use regression instead of traditional statistical methods [3].

Linear regression is a straightforward approach to supervised learning. It considers that the reliance of Y on X1, X2, ... Xp is a linear pattern. But true regression functions are in no way linear, as shown in Graph 9.1.

Although it might appear overly simplistic, linear regression is very useful both theoretically and practically. Let us assume a model $Y = \beta 0 + \beta 1X + e$, where $\beta 0$ and $\beta 1$ are two unknown constants that represent the intercept and slope, also identified as coefficients or parameters, and e is the error term. Given some estimates $\hat{\beta} 0$ and $\hat{\beta} 1$ for the model coefficients, we calculate future sales using $\hat{y} = \hat{\beta} 0 + \hat{\beta} 1x$, where $\hat{y}$ indicates a prediction of Y on the basis of $X = x$. The hat symbol denotes an expected value.

Technically, the linear regression coefficients are determined in order to reduce the error in predicting the outcome of a variable or value. This process of computing the beta coefficients is called the *ordinary least squares* method [4].

Let $\hat{y}_i = \hat{\beta} 0 + \hat{\beta} 1x_i$ be the prediction for Y based on the ith value of X. Then $e_i = y_i - \hat{y}_i$ represents the ith residual.

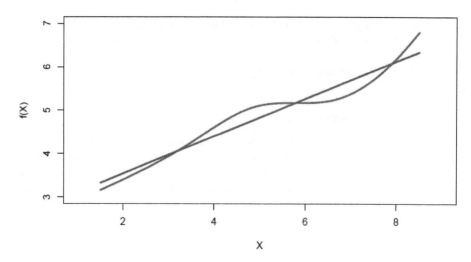

GRAPH 9.1 Linear regression.

When we have multiple predictor values, like X1,X2,X3....XN, then:
our model will be Y = β0 + β1X1 + β2X2 + ··· + βpXp + e, •. We interpret βj as the average effect on Y of a one-unit increase in Xj, holding all other predictors fixed. This concept is known as multiple linear regression.

Sometimes, increasing the value of x1 may also augment the effectiveness of the x2 in explaining the variation in the outcome variable. This is known as interaction effect.

In order to choose an 'optimal' member in the path of models, we use forward or backward stepwise selection. The most direct approach is called as all subsets or best subsets regression: we compute the least squares fit for all possible subsets and then choose between them based on some principle that balances training error with a model size.

In few cases, the connection amid the result and the predictor variables is not linear. In this condition, we must build a *nonlinear regression*.

9.6.1 Application of Regression in Medicine

Multiple linear regression can also be used in analysing future medical expenses. The goal of this analysis is to observe the relationship between different features and to plot a multiple linear regression based on several features such as name, age, physical condition, present medical expenses and so on. These can be used to predict future medical expenses that helps the people to make necessary financial decisions about, for example, insurance if required.

9.7 QUERYING IN MEDICINE

In the past, people obtained information manually from books. Now, everyone uses computers to write and store information. This information is stored in electronic form. Similarly, the uses of computers to store patients' information has become commonplace in hospitals. Medical information of the patient includes various fields

Machine Learning Algorithms in Clinical Decision-Making Systems

FIGURE 9.8 Querying in medicine.

such as name, age, gender, type of disease and so on. This entire information is stored in the form of tables in databases. As the years pass, the patient's data increases year by year, resulting in a wealth of information. Doctors retrieve the useful information from this data based on certain attributes. This helps in proper decision making. The retrieval information is done using queries. A query is a request for the information from a database table. It plays a vital role in retrieving the information from the database. For example, if the doctor wants any information, he or she poses doubts in the form of a query to the database. The database then looks for the related data to the query and sends it to the doctor or user [5]. See Figure 9.8.

In Figure 9.8, encryption of queries takes place before the query is posted to the cloud servers. Similarly, data owners also encrypt their information before sending it to the cloud server. This is done so that the output information is revealed only to the client. This method provides more accurate query support and strong data confidentiality.

9.8 DENSITY ESTIMATION IN MEDICINE

Density estimation is one of the major areas of statistics that contributes to estimating the probability density, based on the observed data. It is the building block of machine learning. In medicine, medical image segmentation is done based on the estimated information. Segmentation is the process of dividing the data into groups containing similar sets of objects in it. Before the density estimation, the first step that a person should perform is a histogram plot. A histogram is the graphical representation of continuous numerical data. First, the data is grouped into bins, and then the number of objects that are falling in that bin are counted. These counts are the frequencies of observation that are plotted on the y-axis and bins are plotted on the x-axis. The number of bins in a histogram play an essential role in controlling the coarseness of the distribution [6].

The two ways of estimating the density are the parametric and nonparametric methods.

1. Parametric method: A particular form of density function (e.g., Gaussian) is assumed to be known and only its parameters $\theta \in \Theta$ needs to be estimated (e.g., mean, covariance).

Sought: p(x, θ);
Performed: {x1, xn} → ^θ.

2. Nonparametric method: No assumptions are made on the form of the distribution. However, many more training examples are required when compared to parametric methods.
Sought: p(x);
Performed: {x1, xn} → p^(x).

The most popular and widely used density estimation method in medicine is the super vector machine (SVM) method.

Super vector machine is a machine learning method that is based on statistical learning. It can solve regression problems, density estimation problems and classification problems.

Consider the following equation, where P(x) is the solution linear operator equation.

$$F(x) = \int_{-\infty}^{x_1} \cdots \int_{-\infty}^{x_N} P(t)dt_1 \cdots dt_N$$

SVM is a nonparametric density estimation approach which is consistent. For solving the density estimation problem using SVM, we introduce an operator A, which is a linear mapping from Hilbert space functions, i.e., from P(x) to F(x) [7].

$$A\,P(x) = F(x)$$

Then, solving the regression problem in image space is done. Moreover, the solutions thus obtained are used to describe image in pre-image space. Then we use SVM, and intensive loss function is applied for density estimation.

9.9 DIMENSION REDUCTION IN MEDICINE

Nowadays a huge amount of medical data is available. This data is available in the form of high dimensions and complex structures, so in order to achieve more accuracy, the irrelevant and inefficient dimensions from the existing dimensions of data has to be removed. This process of reduction of dimensions in the data is known as dimensionality reduction. It a preprocessing step performed to achieve more accuracy and computational efficiency during classification. There are many methods to implement dimension reduction without affecting the classification accuracy. Figure 9.9 shows the categories of dimensionality reduction. The four categories are as follows:

- Feature evaluator and ranker
- Linear dimension reduction algorithms
- Nonlinear dimension reduction algorithms
- Clustering-based feature selection

Machine Learning Algorithms in Clinical Decision-Making Systems

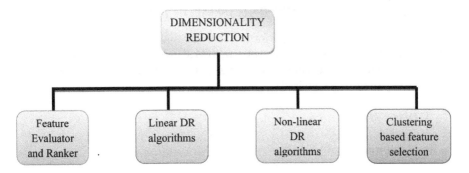

FIGURE 9.9 Categories of dimensionality reduction.

9.9.1 Feature Evaluator and Feature Ranker Algorithm

In order to develop the efficiency and accuracy of the model, several statistical methods and algorithms for feature ranking and feature evaluation are used for selecting the useful and relevant features in the model. These are further classified into filters and wrappers. In a filter, the features are evaluated and then ranked independently. This process uses proxy measures to evaluate the features. In the wrapper, the prediction model is used to assess and evaluate (score) the various features. Error rate is then calculated using the score. Prediction performance in the wrapper is better as compared to the filter.

9.9.2 Linear DR Algorithms

In linear dimensionality reduction, linear combinations of the original features that are enough to project the original data in reduced dimensions are formed. Principal component analysis is widely used to reduce the feature space by obtaining the linear dependencies among various features.

9.9.3 Nonlinear DR Algorithms

Nonlinear dimension reduction is widely used in the area of bioinformatics. It is applied for nonlinear data. Traditional methods for transforming the nonlinear structure into linear dimension space resulted in loss of information. However, methods such as isometric mapping, spectral clustering and Laplacian eigenmaps use special methods in order to retain the lost information of nonlinear structures during transformation.

9.9.4 Clustering Algorithms

Clustering is the grouping of the data based on their similarities. The most broadly used clustering methods are k-means clustering and hierarchical clustering. These methods do not have any prior information available. The methods are explained in Section 9.11.

9.10 TESTING AND MATCHING IN MEDICINE

Medical testing is the process performed in order to diagnose, monitor and detect disease and then establish the course of action. Medical tests are related to molecular diagnostics and clinical chemistry and are usually carried in a medical laboratory.

A variety of tests are advised:

  i. Diagnostic
 ii. Screening
iii. Monitoring

9.10.1 Diagnostic

Diagnostic testing is a process performed to substantiate or conclude the presence of disease in a patient suspecting of possessing a disease. Figure 9.10 shows the diagnosis process.

Following is an example:

- Whenever a patient is suspected of having lymphoma, the diagnostic test called nuclear medicine is used.
- To check the bacterial infection in patient with a high fever, a complete blood count is used.

9.10.2 Screening

Screening tests are useful in the individuals with an elevated risk for a disease to occur. This test is carried out to manage epidemiology, to monitor disease prevalence, to prevent the disease or for strictly statistical purposes.

Following are some examples:

- Measuring blood pressure of a patient based on existing data.
- Measuring blood sugar levels of a patient suspected to be diabetic.
- Measuring the TSH in the blood of a newborn for congenital hypothyroidism.

FIGURE 9.10 Diagnosis process.

Machine Learning Algorithms in Clinical Decision-Making Systems   **179**

9.10.3 MONITORING

Monitoring is a medical test employed to monitor the advancement or response to a medical treatment.

9.10.4 PRECISION OR ACCURACY ASPECTS

The precision of a test is its correspondence with the exact value. Accuracy of the experiment is its reproducible nature when it is repeated on the same set of the sample.

9.11 CLASSIFICATION AND CLUSTERING IN MEDICINE

As we studied in Section 9.2, medical imaging is playing a vital role in the diagnosis of disease. The proper identification or detection of the disease is possible only when you cluster and classify the image data properly. This can be achieved by selecting proper and efficient classification and clustering algorithms [8].

9.11.1 CLUSTERING

Clustering is an unsupervised machine learning method. It is a process of grouping the data based on the similarity constraints. Here, the groups are called clusters. The data or objects in a group are different when compared to the objects in another group [9].

There are four types of clustering:

1. Hierarchical clustering
2. Partition methods
3. Density-based clustering
4. K-NN (K-nearest neighbour)

9.11.1.1 Hierarchical Clustering

Hierarchical clustering is a technique in which a cluster hierarchy called a dendrogram is built. Each cluster in this technique has child clusters, sibling clusters and its common parent. This technique is further divided into two methods [9]:

a. Agglomerative clustering
b. Divisive clustering

Figure 9.11 shows the hierarchical clustering with dendrogram.

9.11.1.1.1 Agglomerative Clustering

Agglomerative clustering initiates with clusters having a single point and repeatedly unites two or more points, forming suitable clusters. Then, computation of all pairs' pattern-pattern similarity coefficients takes place. Every pattern is placed in its own class based on similarity. Later, the merging of the two most similar clusters into one new cluster is carried out, followed by recomputation of inter-cluster

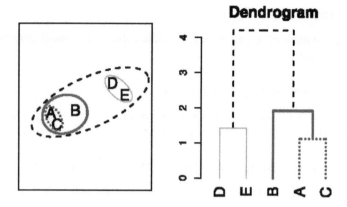

FIGURE 9.11 Hierarchical clustering.

similarity scores. This merging of similar clusters is repeated until k-clusters are formed.

9.11.1.1.2 Divisive Clustering

This method is completely the reverse of agglomerative clustering. Here, the clustering begins with one cluster having the entire points in it. Then the cluster is split into two or more appropriate clusters. The splitting starts from the top, having all patterns in one cluster end up with each pattern in its own cluster.

These methods make it easier for decision making by looking at the dendrogram as shown in Figure 9.11. But these are not applicable for large datasets and are very sensitive towards outliers.

9.11.1.2 Partition Methods

These methods divide the entire database into partitions of k-clusters. It generally uses the iterative optimisation mechanism [9]; that is, iterative reassigning of points between the clusters. These algorithms are classified as:

1. K-means
2. K-medoids

9.11.1.2.1 K-Means Clustering

In k-means clustering, data is partitioned into k number of groups. These groups are disjoint clusters. It consists of two phases. In first phase, the centroid is calculated, and in second phase, points are taken in each cluster which are near to the centroid. This method is most commonly used in health services. For example, people with heart disease are grouped based on their blood pressure and cholesterol levels. K-means clustering is easy to implement and produces tighter clusters than hierarchical clusters. However, it is tricky to expect the number of clusters and it is sensitive to scale.

Machine Learning Algorithms in Clinical Decision-Making Systems

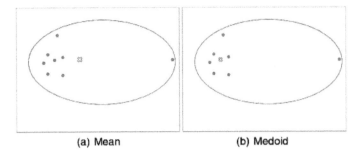

(a) Mean (b) Medoid

FIGURE 9.12 Partition methods of clustering (a) K-means clustering (b) K-medoids clustering.

9.11.1.2.2 K-Medoids Clustering

K-medoids is analogous to k-means, but in k-medoids data points are taken as centres, whereas in k-means it is not necessary that the centre of a cluster is data point; refer Figure 9.12. These methods aim at reducing the distance between the data points.

9.11.1.3 Density-Based Clustering

It is a technique where clusters are the dense areas that are separated from one another by sparse areas. Each cluster has two parameters, i.e. epsilon and minimum points, and each cluster has to contain at least a single point. The number of neighbour clusters must be greater than or equal to the minimum data points. Subsequently, the algorithm continues to iterate for the residual data points in the set. It is robust to outliers and there is no necessity to specify the number of clusters as in k-means. But it does not give good results when the differences in densities are large. Figure 9.13 shows density-based clustering.

9.11.1.4 K-Nearest Neighbour

The K-NN classification rule is to assign to a test sample the majority category label of its k nearest training samples. K-NN assumes that all instances are points in some n-dimensional space and defines neighbours in terms of distance (usually Euclidean

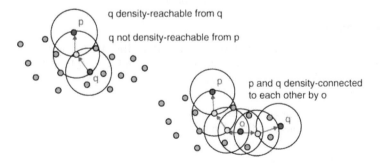

FIGURE 9.13 Density-based clustering.

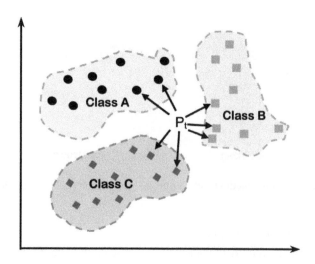

FIGURE 9.14 K-NN clustering.

in R-space). It is a simple technique that is easily implemented, and the cost of the learning process is zero. But it is computationally expensive to find the k-nearest neighbours when the dataset is very large. Figure 9.14 shows the K-NN clustering.

9.11.2 Classification

Classification is the final and major step in the process of diagnosis. It is a supervised learning method where the input data is classified using the existing training data. There are many methods by which the given input data can be classified [10]. These methods are mainly divided into three categories.

9.11.2.1 Texture classification

This classification technique is the key component in many medical applications. It aims at assigning a strange sample data to one of the sets of the known texture classes. It belongs to the texture analysis domain.

9.11.2.2 Neural Networks

It is the most commonly used method in artificial intelligence for problem solving. It has three layers: input, hidden and output, as shown in Figure 9.15.

It mainly consists of five basic steps:

- Finding net output at different output layer nodes.
- Designing an objective function mathematically.
- Designing the cost function.
- Minimising the error.
- Updating the weights.

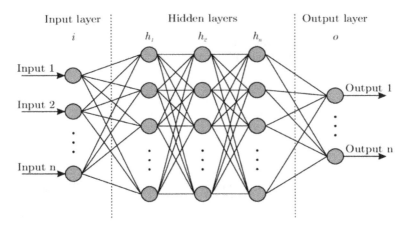

FIGURE 9.15 Neural networks.

9.11.2.3 Data Mining

Classification using data mining techniques involves the usage of sophisticated tools of data analysis to find out the relationships in large volumes of data. The data mining classification techniques that are used for medical imaging classification are as follows:

　i. Modified K-NN: This uses an instance-weighing scheme which is based on distance measure. It gives better accuracy compared to other algorithms.
　ii. One-class KCPA: This method identifies the data objects of one specific class among all other data objects.
　iii. Cascading of one-class KCPA: This algorithm ensures high accuracy. It uses a reject option in order to reduce the cost of misclassification.

9.12 CONCLUSION

Machine learning has changed the entire healthcare system by providing many advantages to the public. It has had a great impact on lifestyles and well-being. It has bridged the gap between patients and doctors, at the same time saving the time and reducing the costs of health services. Its multidisciplinary nature helps in solving health problems using its various disciplines. The improvement of machine learning can lead to the invention of robots that assist doctors during complex surgeries. Mining the data to discover the useful information from the data can also be done, which would enhance improvements in screening and profiling for better drug discovery. This chapter has focused on many new advancements taking place in the field of machine learning that can in turn help in providing better healthcare services to patients.

REFERENCES

1. R. Samant, and S. Rao, A study on feature selection methods in medical decision support systems, International Journal of Engineering Research & Technology (IJERT), Vol. 2, Issue 11, November 2013.

2. K. Rajeswari, V. Vaithiyanathan, and S. V. Pede, Feature selection for classification in medical data mining, International Journal of Emerging Trends and Technologies in Computer Science, Vol. 2, April 2013.
3. G. B. Fauet, and H. C. Davis, Regression analysis in medical research, Southern Medical Journal.
4. C. Combes, F. Kadri, and S. Chaabane, Predicting hospital length of stay using regression models: application to emergency department, HAL Id: hal-01081557, https://hal.archives-ouvertes.fr/hal-01081557, Submitted on 9 Nov 2014.
5. N. Qamar, Y. Yang, A. Nádas, and Z. Liu, Querying Medical Datasets While Preserving Privacy, Procedia Computer Science, Dec 2016, 98, 324–331, 10.1016/j.procs.2016.09.049.
6. L. Diao, H. Yan, F. Li et al., The research of query expansion based on medical terms reweighting in medical information retrieval, EURASIP Journal on Wireless Communications and Networking, 105, 2018,https://doi.org/10.1186/s13638-018-1124-3.
7. M. Kalekar, and B. Sonawane, A survey on medical image classification techniques, International Journal of Innovative Research in Computer and Communication Engineering, Vol. 5, Issue 7, July 2017.
8. Y. K. Alapati, Combining clustering with classification: a technique to improve classification accuracy, International Journal of Computer Science Engineering, Vol. 5, Issue 6, Nov 2016.
9. N. S. Nithya, K. Duraiswamy, and P. Gomathy, A survey on clustering techniques in medical diagnosis, International Journal of Computer Science Trends and Technology (IJCST), Vol. 1, Issue 2, Nov-Dec 2013.
10. R. Sharma, S. Narayan, and S. Khatri, Data Mining Using Different Classification and Clustering Techniques: A Critical Survey, IEEE, 2016.

10 Technique of Receiving Data from Medical Devices to Create Electronic Medical Records Database

Vu Duy Hai
Hanoi University of Science and Technology
Hanoi, Vietnam

CONTENTS

10.1 Introduction 186
10.2 Automatic Technique of Acquiring Analog Electronic Medical Images 186
    10.2.1 Introduction and Classification 186
    10.2.2 Automatic Acquisition of Analog Electronic Medical Image from
           Imaging Equipment with Standard Communication Interface 187
    10.2.3 Evaluation of the Quality of Electronic Medical Image
           after Acquisition 192
10.3 Automatic Technique of Acquiring Digital Electronic Medical Image 194
    10.3.1 Introduction and Classification 194
    10.3.2 Automated Technique of Digital Image Acquisition According
           to DICOM Standard 194
    10.3.3 Evaluation of the Quality of Image Data after Acquisition 198
10.4 Automatic Technique of Acquiring Medical Data in Text 199
    10.4.1 Technical Data Analysis at the Output of Medical Equipment 199
    10.4.2 Building Software to Automatically Receive Electronic Medical
           Data in Text 202
    10.4.3 Evaluation of the Quality of Data after Acquisition 204
10.5 Automatic Technique of Acquiring Medical Data in Graphic 204
    10.5.1 Determine in Detail the Data Structure in Packets at the Output
           of the Device 204
    10.5.2 Technology of Connecting and Automatically Acquiring
           Medical Data in Graph 206
    10.5.3 Evaluation of the Quality of Data after Acquisition 211
10.6 Conclusion 211
Acknowledgements 212
References 213

10.1 INTRODUCTION

In the electronic medical record processing model, the process of digitising medical data is an important key to creating an electronic data source to deploy intelligent IT applications as well as to evaluate the optimisation of the model. Meanwhile, patients' medical data is diverse and complex, with many different types of data, from different types of medical devices, from various manufacturers. Thus, the process of digitising these medical data is an enormous task. In addition, the strict requirements of the integrity, accuracy and confidence level when the patient data is electronically processed is also a challenge [1].

Imaging data from diagnostic devices comply with Digital Imaging and Communications in Medicine (DICOM), while modern laboratories use data management systems such as Radiology Information System (RIS), Picture Archive and Communication System (PACS) or Laboratory Information System (LIS). Therefore, the electronic process of medical data is quite simple and accurate. However, other forms of medical data such as videos, graphics, numerals and others still must be entered manually from a computer keyboard and specialised scanners are used to digitise data from film, and then printed on paper or stored separately on CD and DVD drives [2]. The digitising of medical data using the above methods has shown many shortcomings. According to a study, on average, in Australia about 13.3% of the doctors' conclusions had at least one error caused by typing from a computer keyboard. In New Zealand the average is 50%, the US is 59%, Canada is 60%, Ireland is 65.5% and Sweden is 66% [3]. In developing countries, a major feature is the numerous models of general hospitals with a variety of types and origins of medical devices that generate medical data. Therefore, the process of digitising this data source is even more complicated. This chapter presents several techniques to collect data from different types of medical devices and evaluate the quality of data after receiving it to form a database of I-EMR in an electronic hospital model [4–5].

10.2 AUTOMATIC TECHNIQUE OF ACQUIRING ANALOG ELECTRONIC MEDICAL IMAGES

10.2.1 INTRODUCTION AND CLASSIFICATION

Medical data in the form of images, video and audio are mainly generated from the current diagnostic imaging devices such as diagnostic ultrasound equipment, endoscopic equipment or electron microscopes. These are devices that are considered routine in the health sector during medical examination and treatment. Therefore, the number of medical records of this type created daily in health facilities is very large. However, the manufacturers of this device group only provide data processing functions directly on the device such as display, printing, analysis and temporary storage in the device's memory, CD/DVD storage or film printing. To provide the function of communicating with external peripherals for data processing purposes, the devices are mainly designed in the analog video output port. Therefore, in order to be able to receive and store these types of medical data in the form of electronic medical records on computers, a method of collecting and converting these data into digital forms is required [6–8].

Receiving Data from Medical Devices to Create Electronic Records 187

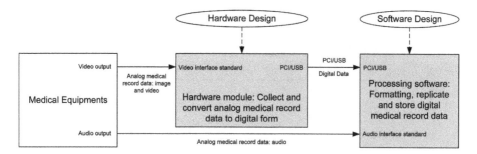

FIGURE 10.1 The hardware module automatically captures the analog electronic medical image from the imaging device with standard communication interface to convert to digital format on the computer.

10.2.2 Automatic Acquisition of Analog Electronic Medical Image from Imaging Equipment with Standard Communication Interface

The principal diagram for the process of automatically collecting analog electronic medical data from imaging equipment with standard communication is shown in Figure 10.1. Accordingly, analog electronic medical data including images, video and audio from imaging devices with communication standards will be collected and converted into digital data by the hardware circuit module. This digital data is then sent to computers and processing software to perform the process of formatting, reproducing, displaying and storing medical data into the database in a digitised form [9].

The technique to implement the proposed process includes two main contents: (1) Designing hardware circuit to capture and convert electronic medical data in the form of analog images and videos into digital form and (2) building computer processing software to format, reproduce and store electronic medical data in the form of digital images and videos.

- *Hardware circuit design captures and converts electronic medical data in the form of analog images and videos into digital form.*

To be able to automatically receive electronic medical data in the form of analog images and videos, first it is necessary to convert these data from analog to digital format with acceptable errors in the processing of medical information. Based on the characteristics of the analog electronic image data and video data in communication standards on medical imaging devices (including horizontal scan frequency, vertical scan frequency, video signal bandwidth, frame rate), this function is proposed to be carried out on electronic circuits with detailed block diagram described in Figure 10.2. In particular, the analog-to-digital convertor (ADC) continuously performs the conversion of analog video signals to digital signals. To ensure the ability to accurately reproduce video signals after digitisation, the sampling frequency selected in the high range is 15 Msps. This sampling rate is greater than the data transfer rate via universal serial bus (USB), so a high-speed 256-kB medium memory has been used to store data temporarily. Accordingly, the lines of continuous output from the ADC are stored in the intermediate memory one by one. The storage process is paused when a complete frame is encoded to begin the data transfer process. The pause is

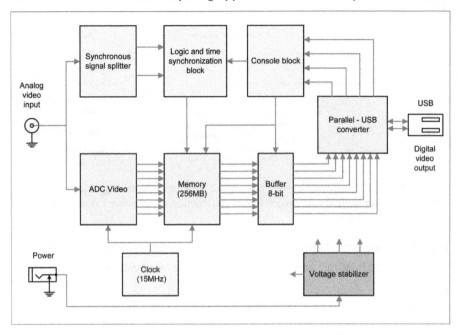

FIGURE 10.2 Circuit diagram proposed to receive and convert electronic video medical data format into digital format at a rate of 3 to 12 frames per second.

maintained until all frame data in the memory is transferred to the computer software via USB [10].

On the other hand, an analog video signal is sent to the synchronous pulse detector. The output of the synchronous pulse detector is a line and frame synchronisation signal that is fed into the logic block and the time synchronisation to ensure that the storage process on the intermediate memory is performed correctly and without overlapping. This block ensures that the archiving is carried out line by pixel until the end of the frame. In addition, control signals received from the USB port sent by the software on the computer are fed into the console block to affect other blocks in the system. This block only affects memory block, logic block and time synchronisation, and 8-bit buffer. Thus, the operation of the ADC block, the synchronous pulse separator and the clock generator are continuous and relatively independent of each other. The detailed design diagram for the acquisition and conversion of analog electronic video medical data into digital format is proposed as shown in Figure 10.3. The parameters to achieve by following design circuit are as follows: inputs are electronic medical data in the form of analog images and videos taken from the composite video plug standard; the output is digitised data and communicates according to USB standard to the computer; conversion speed is from 3 to 12 frames per second; the maximum size of each frame is 730×288 pixels; the encoder resolution is 8 bits per pixel; colour decoding supports Phase Alternating Line (PAL) and National Television System Committee (NTSC) colour systems for processing software.

Receiving Data from Medical Devices to Create Electronic Records 189

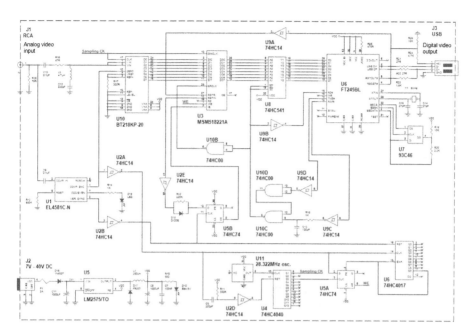

FIGURE 10.3 Detailed design diagram of the receiving and converting electronic medical data from analog video to digital format.

- *Building computer processing software to format, reproduce and store electronic medical records in the form of digital images and videos*

Computer processing software will perform the function of receiving digital data stream from the hardware circuit, then format, reproduce and store it as electronic medical data in the form of digital images and videos. This digital medical record data can be stored as a single image and as a video. Some the main algorithm diagrams and processing functions used to receive, reproduce, store and display electronic medical data in the form of digital images and videos in the processing software are designed as follows.

1. Building algorithm diagram of data acquisition and digital image and video reproduction: The diagram of the acquisition, processing and reproduction of image and video data is described as shown in Figure 10.4. Accordingly, the digital data received from the hardware circuit sent through the USB port will be put into the buffer memory, the software will perform entropy decoding, quantisation decoding, reverse discrete cosine transform (DCT), aggregate the image and finally display the video data [9].
2. Develop an algorithm to save digital image data into a database: The algorithm for saving digital medical image into a database is as described in Figure 10.5. The medical imaging data after being re-created from the hardware circuit will be stored in a temporary folder defined on the computer. This entire image data is then converted to binary bytes so it can be stored in the database. Finally, the image data in the temporary directory will be deleted to free up memory.

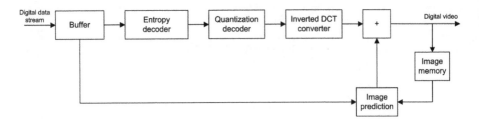

FIGURE 10.4 Schematic of acquisition and reproduction of video medical data from digital stream of hardware circuit.

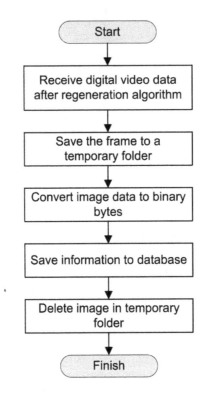

FIGURE 10.5 Algorithm flowchart archives medical data in the form of images and videos into a database on a computer.

3. Develop an algorithm for reading and displaying images from a database: The algorithm for reading and displaying image medical data stored in a database as a binary byte is described in Figure 10.6. The binary byte data stored in the database is read and checked in turn to be the last byte of the image. If not, the binary byte data will be converted to bitmap data and transferred to the Picture Box image data display tool. When the last byte is reached, the bitmap data in the Picture Box tool will be displayed as a digital image.

Receiving Data from Medical Devices to Create Electronic Records 191

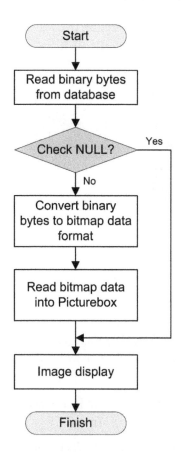

FIGURE 10.6 Algorithm flowchart reads and displays medical data in the form of images and videos from the database.

4. Following are the main processing functions:

```
AxVideoCap1.Device();/function receives data from hardware
circuit posts/
AxVideoCap1. VideoFormat, for example RGB24(800x600);//function
selects data format for image and video posts//
AxVideoCap1.VideoStandard, for example PAL standard or NTSC
standard;//function determines the video standard post//
AxVideoCap1.Start();//function starts to receive and recreate
video data//
AxVideoCap1.Pause();//function pauses the process of receiving
and recreating//
AxVideoCap1.Stop();//function finishes getting the process//
AxVideoCap1.SnapShot2Picture();//function receives and saves
each photo frame//
Dim fs AsNewFileStream (Trim (pathfile,FileMode.Open);//
function of converting image data to bit stream//
DimData() AsByte = New [Byte] (fs.Length);//function of
converting bit stream to byte array//
```

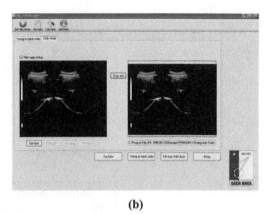

(a) (b)

FIGURE 10.7 The result of an automated process of obtaining analog electronic medical data in image and video type from diagnostic ultrasound equipment. (a) Connect and receive data from Siemens ACUSON X300 and (b) the software interface displays the post-acquisition and processing results.

The automated process of obtaining electronic medical data in the form of analog images and videos from diagnostic imaging devices has been designed and built by the author using the technique described above as illustrated in Figure 10.7.

10.2.3 Evaluation of the Quality of Electronic Medical Image after Acquisition

The author has collected image and video data from several diagnostic ultrasound devices that are being commonly used in hospitals, along with the original data stored temporarily on devices for obtaining evaluation and analysis data. Due to the characteristics of medical data in the form of images and videos including grayscale type (2D black-and-white ultrasound) and colour type (3D, 4D, endoscopy, electron microscopes), it is important to assess the quality of data collected for both types of images.

For grayscale image data, the acquisition and evaluation of three 2D black and white diagnostic ultrasound equipment, each device randomly captures 20 images of 20 different patients. The equipment performing the assessment includes: Logig100-GE device, SSD1000-Aloka device and AcusonX300-Siemens device. Parameters received for grayscale images from these devices are 640×480 pixels image size, 8 bits encoding. Performing the calculation of peak signal-to-noise ratio (PSNR) values for the corresponding image pairs, we are graphed showing the trend of variation of PSNR values described in Graph 10.1 [10, 11].

For colour image and video, acquisition and evaluation on two popular colour diagnostic ultrasound devices, Accuvix XQ-Medison and EUB-6000-Hitachi. These are devices capable of producing colour ultrasound images as well as 3D colour videos when performing ultrasound. On each device, the author randomly collected five colour videos of five different patients. Each video, author evaluates to 20 frames, respectively. The

Receiving Data from Medical Devices to Create Electronic Records 193

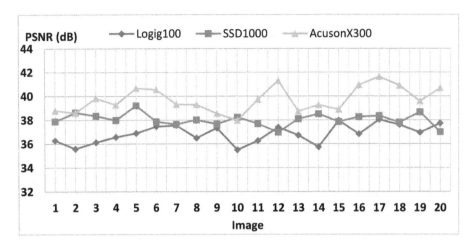

GRAPH 10.1 The chart shows the trend of changing the PSNR value for 20 pairs of multi-level grey ultrasound image data received on the computer and the original image data on the temporary memory of three survey devices: Logig100, SSD1000 and AcusonX300. In the 20 pairs of images assessed, the PSNR value varied from 35.49 dB to 41.63 dB.

received colour video data parameters from these devices include: frame size 640×480 pixels, 24 bits encoding for three colours RGB. The PSNR values for each pair of frames on the two original colour videos from the device's temporary memory and the colour video captured on the computer by the proposed method are shown in Graph 10.2.

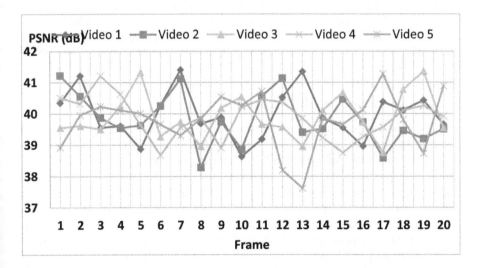

GRAPH 10.2 The graph shows the trend of fluctuating PSNR values for 20 pairs of five colour video frames received on the computer and original video data received from the temporary memory of EUB6000 3D colour ultrasound equipment. In 20 pairs of rated frames, PSNR values varied from 37.62 dB to 41.40 dB.

From the charts showing the trend of fluctuating PSNR values for data acquisition and quality assessment of medical data in the form of images and videos we see: For the Logig100 device, the value of PSNR varies from 35.49 dB to 37.96 dB, the average is 36.84 dB. For SSD1000 devices, the value of PSNR varies from 36.97 dB to 39.21 dB, the average is 38.03 dB. For the AcusonX300, the value of PSNR varies from 37.96 dB to 41.63 dB, the average is 39.72 dB. For Accuvix XQ devices, the value of PSNR varies from 38.19 dB to 41.82 dB, the average is 39.93 dB. For EUB6000, the value of PSNR varies from 37.62 dB to 41.40 dB, the average is 39.87 dB. Thus, compared to the assessment criteria of image and video data quality in processing health information at a good level, PSNR ranges from 35 to 40 dB [11, 12], the method of obtaining analog electronic medical data in the form of images and videos presented above is completely responsive. It can be used to automate the process of digitising of analog image and video medical data from medical imaging devices.

10.3 AUTOMATIC TECHNIQUE OF ACQUIRING DIGITAL ELECTRONIC MEDICAL IMAGE

10.3.1 INTRODUCTION AND CLASSIFICATION

Medical data in DICOM image format, graphic and text format (alphanumeric) are usually generated from medical devices such as CT scanners, MRI, DR, functional diagnostic equipment and laboratory machines. The patient's data, after being measured from the patient, is usually temporarily stored in the memory of the device in digital form. In addition, most of these devices' manufacturers often have built-in digital communication standards to connect with peripheral devices. For current DICOM image devices, most of the output of the device is designed with DICOM standard for digital image data communication with RJ45 connection. For devices creating data in graphical or alphanumeric format, the output of the device has standard RS232 digital communication. This section presents methods for automatically collecting digital electronic medical data from digital communication outputs on medical devices [10].

10.3.2 AUTOMATED TECHNIQUE OF DIGITAL IMAGE ACQUISITION ACCORDING TO DICOM STANDARD

DICOM is a world standard that sets the rules for formatting and exchanging digital imaging data and patient-related information in health. Digital image data generated from different devices following DICOM standards will create a common 'language', allowing the process of communicating images and related health information between applications to be done easily. DICOM is an open standard, and it exists through the procedures of the DICOM standards development committee. DICOM is based on a realistic model with patient information, imaging data and reporting data. These models are called entity relationship models (E-R). The approach to developing data structures based on an entity relational model is called object-oriented design (OOD). Objects are entities that are defined by the model. The characteristics of each entity are called attributes. DICOM calls model-based objects information objects (IO). Models and attribute tables define them as information object definitions (IODs). The entities in the model are abstract. If an actual value replaces a property, that entity is called an instance [13].

Receiving Data from Medical Devices to Create Electronic Records 195

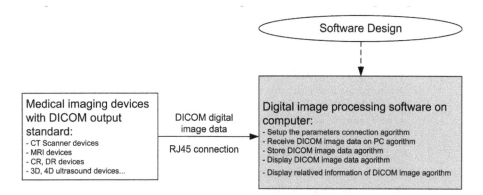

FIGURE 10.8 Process flow diagram automatically captures digital image data according to DICOM standards.

Figure 10.8 shows a process diagram of automatic acquisition of digital image data according to DICOM standard. Digital data at the output of the medical imaging device will be sent to the computer via the RJ45 connector. Through the services in DICOM standard, the computer processing software will receive and store image data into the database [9, 10].

- *Building software to process image data according to DICOM standard on the computer:* The function of the data processing software on the computer is to automatically receive digital data from imaging devices sent through the RJ45 connection port according to DICOM standard. Through services in the DICOM standard, this digital image data is re-created and stored into a database. It also provides display and image processing features according to DICOM standards in electronic medical model applications. With the DICOM Objects tool, image data processing according to the DICOM standard is performed by the following algorithms and processing functions:

  - *Building algorithm to set parameters for connecting with imaging equipment.* The library in DICOM Objects supports the DICOM standard C-STORE service, Therefore, in order to send images from an imaging device to an external device via the DICOM standard, a setup window must first be built to enter the IP address and port of the service class provider (SCP) image, set up application entity title (AET) of SCP and service class user (SCU). The algorithm of setting parameters connecting with imaging devices is implemented in the dot.NET environment as follows:

  Dim im As DicomImage;
  Dim res As Integer;
  im = Viewer.Images;
  res = im.Send(IP address and image port of SCP, AET of SCU, AET of SCP);

  - *Building algorithm to receive DICOM images on computers.* All DICOM Viewer objects in DICOM Objects have the ability to 'listen' to the signal to send images from imaging devices. When an image signal is sent, DICOM Viewer will receive the data transmitted from SCP. The image receiving algorithm is implemented in the dot.NET environment as follows:

Dim server As New DicomServer;
server = New DicomServer;
server.Listen(the listening port);
DicomServer.

- *Building algorithm to save DICOM images into the database.* Once DICOM images have been sent from the device to the computer, the DICOM image storage needs to be performed to write the data received to the storage drive, simultaneously reading file information and writing to the database for later retrieval and retrieval of data. The process of processing and recording files is performed as follows. Create a path that is expected to write the image file to the database. Here the image file name is taken using the labelled component (0008,0018). Extract the information in each image file and then push the information and the path of image recording into the DICOM database for searching and retrieving images. Use the M-WRITE service in DICOM standard to create new files in a file set and assign them a file index or update files if the existing file set already exists. The database for storing DICOM file information includes the components described in Table 10.1.

TABLE 10.1
Components Used to Store DICOM Image Information

No	Data Component Label	Data Component Description
1	(0008,0020)	Study Date
2	(0008,0021)	Series Date
3	(0008,0030)	Study Time
4	(0008,0031)	Series Time
5	(0008,0060)	Modality
6	(0008,1030)	Study Description
7	(0008,103E)	Series Description
8	(0008,1048)	Physician of Record
9	(0010,0010)	Patient's Name
10	(0010,0020)	Patient's ID
11	(0010,0030)	Patient's Birth Date
12	(0010,0040)	Patient's Sex
13	(0020,000D)	Study Instance UID
14	0020,000E	Series Instance UID
15	(0020,0010)	Study ID
16	(0020,0011)	Series Number
17	(0020,0013)	Instance Number

The M-WRITE service in DICOMObjects will support the implementation of writing DICOM image data to the drive according to the previously created path. The algorithm for recording image data on dot. NET environment is as follows:

ImageReceived.Write(filename, "1.2.840.10008.1.2.1");

Where: « filename » is the index created, "1.2.840.10008.1.2.1" is the conversion syntax for existing VR and little-endian formats. Can be replaced by other conversion syntax. DICOMObjects supports all conversion syntax according to the DICOM standard defined.

- *Building algorithm to display DICOM images on computers.* The display of image data and related information in the DICOM file is the process of reading the data components in the DICOM file through decoding and displaying it on the screen. The diagram depicting the display of images and related information in the DICOM file is done according to the steps described in Figure 10.9.

  In the DICOMObjects tool, there is a DICOMViewer object that allows decoding data according to the image encoding standards in DICOM to display. Functions that control the display of image data include:

  DicomViewer.Images.Read (filename of images in local directory);
  DicomViewer.Images.Add (e.Instanceis the image received or retrieved from the server);
  Display multiple images at the same time:
  DicomViewer.Multicolums=n; (n is the number of columns);
  DicomViewer.Multirows=m; (m is the number of rows).

- *Building algorithm to display related information in DICOM file.* To display related information, DICOMObjects supports DICOM Labels object. So

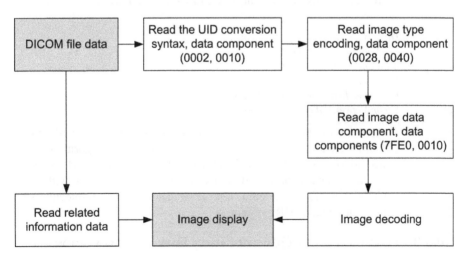

FIGURE 10.9 Diagram depicting the process of displaying image data and related information in a DICOM file on a computer.

TABLE 10.2
Information Related to DICOM Images

No	Related Information	Data Component
1	Model Name	(0008,1090)
2	Study ID	(0020,0010)
3	Series Description	(0008,103E)
4	Current Series	(0020,0011)
5	Current image/total number of images	(0020,0013)/(0020,1002)
6	Number of Rows	(0028,0010)
7	Number of Columns	(0028,0011)
8	Patient's Name	(0010,0010)
9	Name of Institute	(0008,0080)
10	Patient's ID	(0010,0020)
11	Study Date	(0008,0022)
12	Series Time	(0008,0032)
13	Image Size	(0018,1100)
14	Thickness	(0018,0050)
15	Brightness	(0028,1050)
16	Contrast	(0028,1051)
17	Specifications related to the study process	(0018,0091), (0018,0080), (0018,0081), (0018,1250)

read the information in the image and assign it to the DICOM Labels object. In fact, information related to DICOM images is often displayed with image data including the information described in Table 10.2 as follows.

To access this component, DICOMViewer supports the following function:

```
DicomViewer.images(group,element);
InfoLabel=New DicomLabel();
InfoLabel.Left= m;
InfoLabel.Height=n;
InfoLabel.Text=DicomViewer.images(group,element);
DicomViewer.Labels.Add(InfoLabel);
```

The process of automatically collecting digital image data according to DICOM standard has been researched and built by the algorithm presented above. Figure 10.10 depicts the results of the automated process of acquiring and displaying digital image data in accordance with the DICOM standard on the built-in processing software by connecting to Airis Mate-Hitachi MRI system.

10.3.3 Evaluation of the Quality of Image Data after Acquisition

Like the method of evaluating the quality of image data received from imaging devices with the analog output interface, the DICOM image data collected on the computer and

Receiving Data from Medical Devices to Create Electronic Records 199

(a)

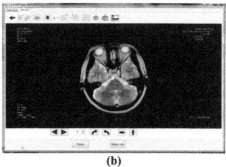
(b)

FIGURE 10.10 The results of the process of automatically acquiring and displaying digital image data in accordance with DICOM standard by connecting with Air is Mate-Hitachi MRI system. (a) DICOM data retrieval result from data base and (b) display DICOM image data and related information on processing software.

the processing software is grayscale. The author used the results of the above software construction, proceeded to connect and receive DICOM medical imaging data on three imaging devices that follow the current popular DICOM standard, including (1) MRI device from Air is Mate-Hitachi; (2) DR device from RADspeed Safire-Shimadzu and (3) CT scanner device from Pronto XE-Hitachi. Each device will randomly receive 20 DICOM images of 20 different patients. Also, the original DICOM image data is from the temporary memory of each device. Parameters received for grayscale DICOM images from these devices include the following: The Air is Mate device has a 256 × 256 pixels image size and 16 bits encoding. RADspeed Safire device has image dimensions of 2544 × 3056 pixels, 12 bits encoding. The Pronto XE device has an image size of 512 × 512 pixels and 13 bits encoding. The trend of fluctuating PSNR values by aggregated results is shown in Graph 10.3 [11, 12].

From the calculation results, the chart shows the trend of changing PSNR values for data collection and evaluation of DICOM medical data quality as follows: For Air is Mate MRI equipment, the value of PSNR varies from 57.26 dB to 61.24 dB, the average is 59.36 dB. For Pronto XE CT scanner, the value of PSNR varies from 53.16 dB to 57.27 dB, the average is 55.48 dB. For RADspeed Safire DR equipment, the value of PSNR varies from 44.25 dB to 47.26 dB, the average is 45.83 dB. Thus, compared to the assessment criteria of image quality data processing in health with a good level, PSNR ranges from 35 to 40 dB for DR images and from 45 to 50 dB for CT and MRI images [11, 12]; the automatic method of obtaining digital image data in DICOM format is to meet the requirements.

10.4 AUTOMATIC TECHNIQUE OF ACQUIRING MEDICAL DATA IN TEXT

10.4.1 TECHNICAL DATA ANALYSIS AT THE OUTPUT OF MEDICAL EQUIPMENT

Medical records in text are most often generated from laboratory equipment such as haematology, biochemistry, microbiology, immune and urine. The test results are usually measurement values, analysing parameters in patient samples. After measurement,

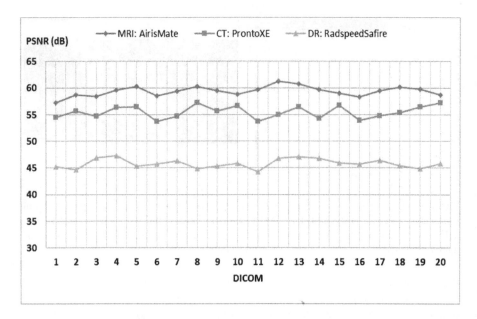

GRAPH 10.3 The graph shows the trend of changing the PSNR value for 20 pairs of DICOM image data collected on the computer and the original DICOM image data received from the temporary memory of three survey devices is Air is Mate, Pronto XE and RADspeed Safire. In the 20 DICOM image pairs evaluated, PSNR values varied from 44.25 dB to 47.26 dB for DR equipment, from 53.16 dB to 61.24 dB for CR and MRI devices.

the results are usually displayed directly on the screen of the device for the doctor to analyse and diagnose. It also provides paper-based printing functions for medical record keeping. Most of these medical data generating devices are designed by manufacturers with digital data communication ports at RS232 output to serve communication with peripheral devices. Because the measurement results are in text, the data of each measurement result will be sent to the output of the device in the form of a message. The structure of this message is usually formatted as described in Figure 10.11 consisting of three segments. In particular, the STX segment is the opening characters of the message, the ETX segment is the characters ending the message and the DATA segment consists of fields of fixed length, containing the characters describing the content of the test results and related parameters. The detailed structure of the DATA segment depends on the type of laboratory device and the manufacturer [14, 15].

Detailed definitions of this structure are based on the principle that device manufacturers must provide in the service manual when selling the product. In practice, however, most manufacturers do not provide or provide these, but the

FIGURE 10.11 Message structure at the output of laboratory equipment.

Receiving Data from Medical Devices to Create Electronic Records 201

facilities are no longer stored. Therefore, in the absence of information on the definition of the message structure sent to output on the device, in order to automatically receive, process and store medical data in text on computer and processing software, it is necessary first to have a method of determining the detailed definition of the message structure at output of the device. The proposed method is to use available software to capture and display digital data at the RS232 port (such as Terminal software, Virtual Serial Ports) of the device as a string of characters. Then it is combined with the results displayed on the device itself for the message sent to the output to analyse, identify and interpolate the rule that defines the data value components in that message.

Figure 10.12a shows the Terminal software interface showing the result of capturing a message from the RS232 port at the output of Bayer's Advia60 haematology device. Figure 10.12b is the corresponding test result shown on the screen of the

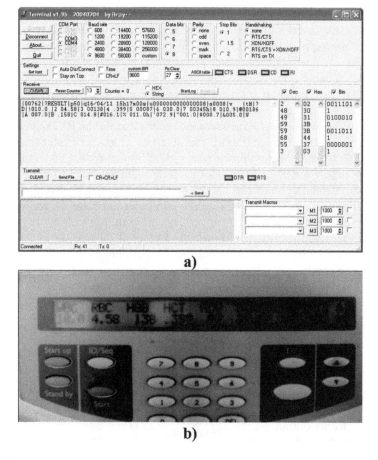

FIGURE 10.12 Illustrating the method of analysing the message structure at the output of the device: **(a)** The result shows the content of the message as a string of characters on Terminal software and **(b)** the results show the parameters on the screen of the device.

device. The comparison between the two results received is the message content in the form of a string on Terminal software and the test results shown on the screen can interpolate the rule that defines the data components in the message as follows: The test order number is 00762, the testing date is 15:17 on 16 April 2011, the test code for the day is 8, the test results: The number of RBC is 4.58, the Haemoglobin is 138, the Haematocrit is 0.399, the MCV is 87, MCH is 30, MCHC is 345, the number of PLT is 186.

By interpolating the format of the message structure received at the output of the devices shown above, the process of automatically acquiring and analysing medical data in text is performed by constructing an algorithm to receive and process data on computer software as described in the following section.

10.4.2 Building Software to Automatically Receive Electronic Medical Data in Text

In order to automatically obtain electronic medical data in text from the laboratory equipment generated, the data connection diagram is shown in Figure 10.13. Accordingly, the test results are sent to the output according to RS232 standard in the form of messages. Based on the message structure format analysed by the above method, the software on the computer will perform analysis and processing of each corresponding data field to display and store on the computer [15].

The algorithm flowchart for the process of automatically receiving and separating data fields in the message received for the processing software is described in Figure 10.14. Accordingly, the message received from the RS232 port will be structurally checked through the two characteristic characters that begin and end the message. If true, the next procedure will be implemented. The data fields in the DATA segment will be identified, extracted and assigned to corresponding variables representing values in patient test data. The result of the assignment will then be displayed and saved into the computer database. In addition, each test result usually sets a normal range. The program algorithm compares and checks the received results. If the measured value is outside the normal range, the algorithm will issue a notice about the content outside the allowed range on the displayed results.

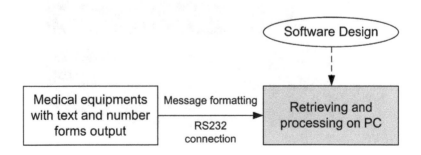

FIGURE 10.13 Connection diagram and medical data in text collection.

Receiving Data from Medical Devices to Create Electronic Records

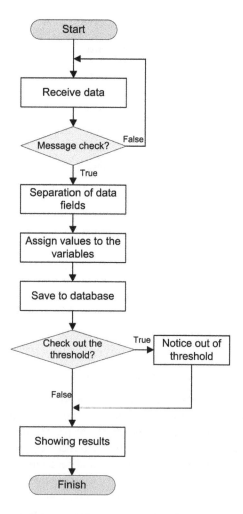

FIGURE 10.14 Algorithm flowchart automatically captures and displays medical data in text at the output of laboratory equipment.

The results of the content of the proposed procedure for automatically collecting medical data in text from laboratory equipment include two subsections: (1) Proposing and implementing the interpolation analysis method to define the message structure received at the RS232 communication output of laboratory equipment and (2) Based on the results of the interpolation method that defines the structure of the proposed message, build software that automatically captures electronic medical data in text through connection with laboratory equipment via the RS22 communication standard. Figure 10.15 illustrates the results of the collection of medical data in text from Bayer's Advia60 haematology equipment.

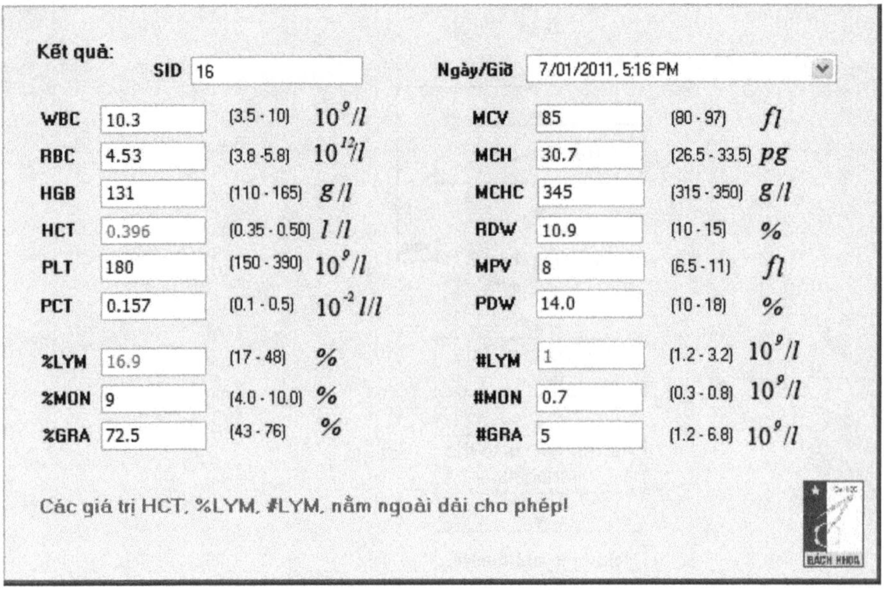

FIGURE 10.15 The results automatically capture and display medical data in text at the output of Bayer's Advia60 haematological laboratory equipment.

10.4.3 EVALUATION OF THE QUALITY OF DATA AFTER ACQUISITION

The purpose of evaluating the results of data collection is to format text on a computer to check the accuracy of the post-acquisition data compared to the original data on medical data generators. Using the results of the process of automatically collecting the above, the author made the connection and received data from three different laboratory machines. Devices for connecting and receiving include: Haematology device KX21-Sysmex, biochemistry device 7150-Hitachi and Urilux-Roche urine device. In each type of device, the author randomly collected the test results of different patients to get analytical and evaluation data. The results showed that, because the data at the output was digital and transmitted at low speed through the RS232 standard with each discrete message each time the device finished a test, the results were received on the computer and there is no difference from the original data. Therefore, with the automatic acquisition method proposed by the author, electronic medical records data in text (laboratory results) after the acquisition is completely identical to the original data from the medical device that created them.

10.5 AUTOMATIC TECHNIQUE OF ACQUIRING MEDICAL DATA IN GRAPHIC

10.5.1 DETERMINE IN DETAIL THE DATA STRUCTURE IN PACKETS AT THE OUTPUT OF THE DEVICE

Waveform or graphical data is usually generated from measuring devices that monitor bio-signals and parameters continuously over time. Typical waveforms are ECG, EEG,

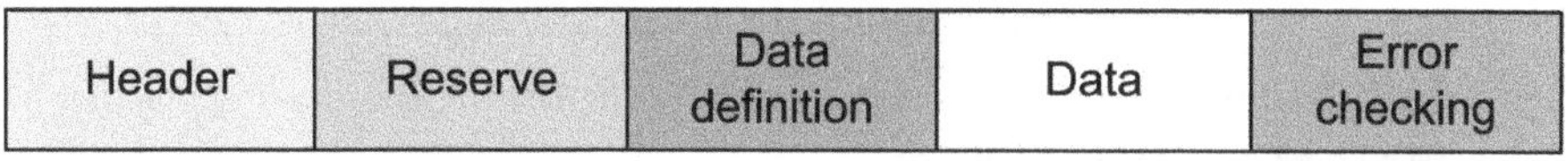

Receiving Data from Medical Devices to Create Electronic Records

Header	Reserve	Data definition	Data	Error checking

FIGURE 10.16 The data frame structure is often used for packet information transmission at the waveform output device.

EMG EOG, ERG, ECoG and RESP signals. During measurement, the results are usually displayed directly on the screen of the device for the doctor to analyse and diagnose. It also provides functions printed on paper to serve the storage of medical records. Like text-based data creation devices, most of these graph-based data creation devices are designed by manufacturers with digital interface at RS232 output. Because bio-signals are continuously measured over time, data of the measurement results will also be sent to the output continuously. However, in order to identify the output digital stream, the data must be sent out for certain packets. Each packet must be defined in a frame with the data structure usually shown in Figure 10.16 as follows [16, 17].

Typical parameters during the data transmission to the RS232 output include: Data bit 8, stop bit 1, parity bit No, bit rate 9600 bauds, data transfer rate 50 packets per second. Similar to medical data in text generation devices, the definition of this data frame is the principle that medical device manufacturers will provide in the accompanying service manual. However, in fact, many manufacturers do not provide enough. Therefore, in the absence of information about the definition of a data frame sent to output on a device, before performing automatic acquisition, reproduction and archiving of waveform data on computers, we must apply the method of determining the detailed definition of the data frame structure in each outgoing packet presented in Section 4.1. An example detailing the definition of MMED6000DP-S6 multi-parameter patient monitoring device data sheet of MMEDCHOICE is as follows [18]:

Each packet will consist of 12 bytes as described in Table 10.3, where 0x55, 0xAA are the signalling bytes at the beginning of the packet. SUM is the error checking byte with the algorithm: SUM = (STATUS0 + STATUS1 + DATA + ECGW3 + ECGW2 + ECGW1 + ECGW0 + SATW + RESPW)/256. STATUS0 is the reserved byte. STATUS1 show definitions of display. Byte DATA contains the data of the parameters as defined in the byte STATUS1. 4 bytes ECGW3, ECGW2,

TABLE 10.3
Transmission Data Protocol (12 bytes)

Pack Head		Reserved	Identification	Identified by the STATUS1		
0x55	0xAA	STATUS0	STATUS1		DATA	
					Respiration Wave	Check Sum
	ECG Wave Sample Point Value			Reserved		
ECGW3	ECGW2	ECGW1	ECGW0	SATW	RESPW	SUM

TABLE 10.4
STATUS1 Structure (1 byte)

STATUS1			BIT7 = 1	ECG beep flag
			BIT6 = x	Reserved
			BIT5 = x	Reserved
			BIT4 = x	The following byte DATA values exceed 255. E.g. if BIT4 = 0 the DATA value is 30, the real value is 30, if BIT4 = 1, the DATA value is still 30, but the real value is 30 + 256 = 286.

BIT3	BIT2	BIT1	BIT0	The following byte DATA means
0	0	0	0	ECGS
0	0	0	1	STAS
0	0	1	0	NIBPS
0	0	1	1	Heart rate (0-255)
0	1	0	0	Pulse rate (0-254) from SpO2
0	1	0	1	Pulse rate (0-254) from NIBP
0	1	1	0	ST
0	1	1	1	% SpO2 (0-99%)
1	0	0	0	Cuff pressure value/2 (mmHg)
1	0	0	1	Systolic 0-255 (mmHg)
1	0	1	0	Diastolic 0-255 (mmHg)
1	0	1	1	Mean arterial 0-255 (mmHg)
1	1	0	0	Respiration rate (0-99)
1	1	0	1	Temperature 1 (T1)
1	1	1	0	Respiration wave gain
1	1	1	1	Temperature 2 (T2)

ECGW1 andECGW0 defines the ECG waveform at one point; ECGW3 is the last value. Byte SATW definition of SpO2 waveform. Byte RESPW defines the respiratory waveform. The structure of the bytes in the packet is described in Tables 10.4 and 10.5.

10.5.2 TECHNOLOGY OF CONNECTING AND AUTOMATICALLY ACQUIRING MEDICAL DATA IN GRAPH

Most waveform or graphic medical data generation devices use the RS232 standard for output data port. Therefore, the data acquisition computer can connect directly with the devices through this connection. The data connection diagram is described

Receiving Data from Medical Devices to Create Electronic Records

TABLE 10.5
DATA Structure (1 byte)

ECGS	The state of ECG: BIT7 = 1 means lead off BIT7 = 0 means normal BIT6 BIT5 mode selection 00: diagnosis mode, bandwidth = 0.05–100 Hz 01: monitor mode, bandwidth = 0.5–75 Hz 10: operation mode, bandwidth = 0.5–25 Hz
STAS	The state of SpO2: BIT7 = 1 means probe off BIT7 = 0 means normal BIT6 = 0 reserved BIT5 = 1 means drop in SpO2 BIT4 = 1 means searching too long BIT3–BIT0 Real time Bar Graph, in the range of 0 to 8, < 3 means weak signal
NIBPS	The state of NIBP: BIT7 = 1 means repeat measurement BIT5, BIT6 reserved BIT3 = 1 means manual BIT3 = 0 means automatic BIT4 = 1 means gas circuit is jam, 0 means normal BIT2, BIT1 mode selection 00: adult mode; 01: paediatric mode; 10: neonate mode BIT0 = 1 means measuring or calculating 0 means not measured or finished
ST	ST value (show as complementally code in the binary system) E.g. –80 means –0.8 mV, 80 means 0.8 mV
T1	Channel 1 temperature value (0–255) The value should divide by 10 then add 20 and then the result is the temperature value E.g. 235 (235/10) + 20 = 43.5°C
T2	Channel 2 temperature value (0–255) same as T1

as shown in Figure 10.17. Graphically electronic medical data is measured over time and sent to the standard RS232 output continuously as packets. These packets will be transmitted to the computer via the COM port. Computer processing software will receive and store it in the form of graphical electronic records. Because medical data

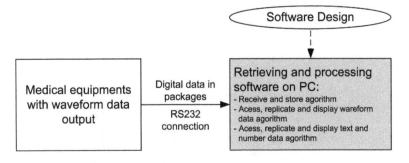

FIGURE 10.17 Data connection diagram in the process of acquiring graphical electronic medical records through standard RS232 connection.

is graphic, the following algorithms need to be developed and applied in processing software design.

- *Building flowchart algorithms to receive and store data into databases:* The algorithm of acquiring and storing graphical data into a database is implemented as described in Figure 10.18. Data from the COM port will be checked through the frame information. If data is available, it is necessary to receive and separate the data components in the packet according to the

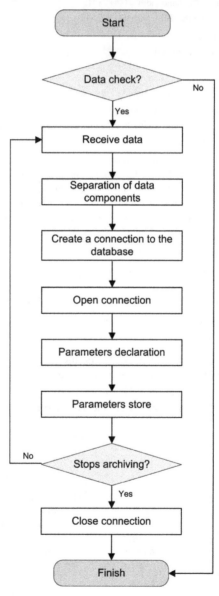

FIGURE 10.18 Algorithm for receiving and storing data received via COM port.

Receiving Data from Medical Devices to Create Electronic Records 209

analysed and defined format structure. Then, create a connection and open the connection to populate the database fields with the declared parameters. Check the archive stop command; if not then continue to receive data. If there is a stop signal, close the connection and end the process of receiving and storing data into the database [16].

- *Building algorithm to retrieve and display data in graphical form:* To retrieve and display graphical data in a database, the algorithm is built as shown in Figure 10.19. Accordingly, data taken from the database will be included in the array of input points. To display a waveform, it is often necessary to select the display mode, on-screen display area and display colours. After the array of points has been formatted according to the selected parameters, the array of points will be converted to the Oxy coordinate system for the graphics program to perform. The waveform will be a set of points drawn according to the formatted input point array.
- *Develop algorithm to retrieve and display alphanumeric data graphically:* The procedure for retrieving and displaying graphical alphanumeric data containing values on measurement of living parameters of patients

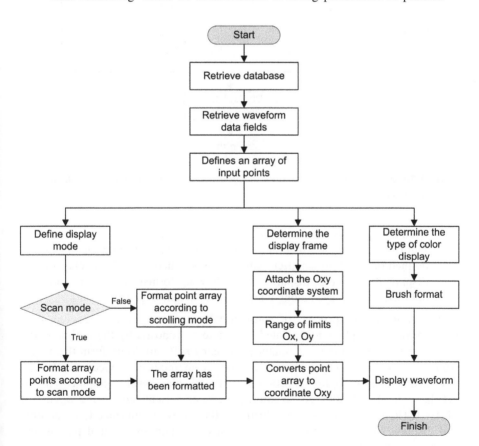

FIGURE 10.19 Algorithm flowchart and display graph data in graphical format.

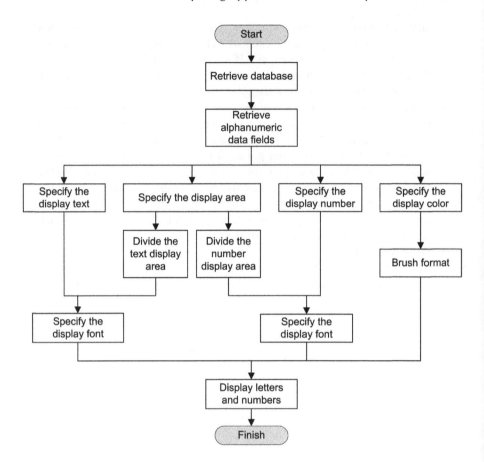

FIGURE 10.20 Algorithm flowchart for the procedure to retrieve and display alphanumeric data in a database.

from a graphical database is built according to the algorithm described in Figure 10.20. The information to be displayed is taken from the data fields defined in the database; after defining the parameters for the display area display, display font, display colour, drawing format, the alphanumeric data will be displayed along with the corresponding graphical data.

The result of the process of automatically acquiring and processing data in the form of graphics and related parameters proposed by the author is the design and construction of computer processing software to perform the acquisition, storage and graphical data display from medical devices via RS232 interface standard. Figure 10.21a shows the display results graphical medical data on MMED6000DP-S6 multi-parameter patient monitoring device of MMEDCHOICE. Figure 10.21b illustrates the software interface that captures and displays on the computer graphical medical data and related parameters when connecting to the device.

Receiving Data from Medical Devices to Create Electronic Records 211

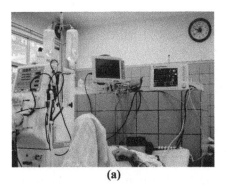
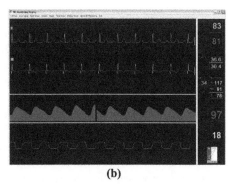

(a)   (b)

FIGURE 10.21 The results of the process automatically receive graphical data and related parameters. (a) Measurement data and display on the MMED6000DP-S6 patient monitoring device and (b) data collected and redisplayed on computer software.

10.5.3 Evaluation of the Quality of Data after Acquisition

Like medical data in text, the purpose of assessing the results of graphical data collection on the computer is to check the accuracy of the post-acquisition data compared to the original data on medical data generators. Using the results of the automated graphical data collection process proposed above, the author made a connection and received data from two different types of waveform medical data generators. Devices that connect and receive are Cardiofax-Nihon Kohden ECG device and MMED6000DP-S6 patient monitor. In each type of device, the author randomly collected the measurement results of many different patients to get analysis and evaluation data. From the aggregate results show that because the data at the output is digital and transmits at low speed through the RS232 standard for each packet, the results received on the computer are not different from the original data. In cases of error from the measuring device (due to slip of the electrode, non-contact measuring electrode or vigorous movement of the patient), then the results obtained will not be determined. In these cases, the measuring device needs to be checked and measured again. Therefore, with the automatic acquisition method proposed by the author, the graphical electronic medical data collected completely matches the original data from the medical device that created them [19].

10.6 CONCLUSION

This chapter focused on the techniques of automatically collecting electronic medical data from various medical devices, implementing collecting and assessing quality of medical data after it is collected on computers. The research results show the following.

For the automatic technique of acquiring analog electronic medical data and assessing the results received from devices with standard communication interface, including analog image and video medical data from imaging devices such as ultrasound, endoscopes, electron microscopes, the author presented a method of designing hardware circuits to receive and convert medical data from analog images and videos into digital form, a method of building data processing software on computers, conducting a system design connection to receive data on five different types of ultrasound machines and assessing the quality of electronic medical data in the form of images and video after receiving. The results show that, for grayscale image, the data quality after acquisition compared to the original data expressed through PSNR values reaches from 35.49 dB to 39.72 dB. For colour image and video, the quality of electronic medical data after receiving compared to the original data expressed through PSNR values reaches from 37.62 dB to 41.82 dB. The requirement for medical information processing for ultrasound image data, PSNR is from 35–40 dB. In conclusion, the process of automatically acquiring analog image and video by the author's proposed method has met the requirements in electronic medical image processing.

For the automatic technique of digital electronic medical data collection and evaluation of the results, including DICOM image data, graphic and alphanumeric data from medical devices such as MRI, CT scanner, DR (DICOM format); ECG, EEG, EMG, patient monitor (graphic format); laboratory machines (alphanumeric), the author presented the method of analysing and interpolating the message structure at the output of laboratory equipment, develop algorithms and software to collect and process medical data on computers, including, DICOM image data acquisition and processing software, software for collecting and processing alphanumeric medical data, graphical medical data. Make connection with three DICOM imaging devices, three laboratory devices and two patient monitors. Evaluation of post-acquisition medical data shows that, for DICOM medical imaging data, the quality of image data after receiving compared to the original image shown in PSNR values reaches from 44.25 dB to 47.26 dB (DR images); from 53.1 dB to 61.24 dB (CT and MRI images). Whereas the required value in processing medical data for DR image is from 35–40 dB, for CT and MRI image it is from 45–50 dB. For graphical and numerical medical data, the evaluation results show that the data after acquisition and the original data are completely matched. This makes sense in overcoming confusion and errors in the process of digitising of alphanumeric data using computer input methods.

As the results obtained show, the techniques and methods proposed by the author for the automated electronic medical data collection process from medical devices on computers have shown the quality of post-received medical data that fully meets health information processing requirements. Therefore, these results can be applied to the model of I-EMR processing system following the trend of automatically collecting data from medical devices.

ACKNOWLEDGEMENTS

The author would like to acknowledge the support of Biomedical Electronics Center, School of Electronics and Telecommunications, Hanoi University of Science and Technology, Bach Mai and Viet Duc hospitals in assisting with connecting and collecting data from medical devices. The author would like to thank Amit J. Nimunkar for help during manuscript preparation.

REFERENCES

1. Asangansi, I. E., Adejoro, O. O., Farri, O., and Makinde, O. (2008). 'Computer use among doctors in Africa: survey of trainees in a Nigerian teaching hospital'. Journal of Health Informatics in Developing Countries. 2: 10–14.
2. Smith, P. D. (2003). 'Implementing an EMR System: one clinic's experience'. Family Practice Management. 10: 37–42.
3. Callen, J. et al. (2009). 'Accuracy of medication documentation in hospital discharge summaries: a retrospective analysis of medication transcription errors in manual and electronic discharge summaries'. International Journal of Medical Informatics. doi:10.1016/j.ijmedinf.2009.09.002.
4. Damera-Venkata, N., Kite, T. D., Geisler, W. S., Evans, B. L., and Bovik, A. C. (2000). 'Image quality assessment based on a degradation model'. IEEE Transactions on Image Processing. 9(4): 636–650.
5. Perera, I. (2009). 'Implementing healthcare information in rural communities in Sri Lanka: a novel approach with mobile communication'. Journal of Health Informatics in Developing Countries. 3(2): 24–29.
6. Kamadjeu, R. M., Tapang, E. M., and Moluh, R. N. (2005). 'Designing and implementing an electronic health record system in primary care practice in sub-Saharan Africa: a case study from Cameroon'. Informatics in Primary Care. 13: 179–86.
7. Hillestad, R., Bigelow, J., Bower, A., Girosi, F., Meili, R., Scoville, R., and Taylor, R. (2005). 'Can Electronic Medical record systems transform health care? Potential health benefits, savings, and costs'. Health Affairs. 24(5): 1103–1117.
8. Mohd-Nor, R. (2011). 'Medical imaging trends and implementation: issues and challenges for developing countries'. Journal of Health Informatics in Developing Countries. 5(1): 89–98.
9. Hai, V. D., Thuan, N. D., and Ngoc, P. P. (2011). 'Automatic retrieving data from medical equipments to create Electronic Medical Record (EMR) for e-Hospital model in Vietnam'. The 5th International Symposium on Bio and Medical Informatics and Cybernetics (BMIC), Orlando, FL, Vol. II, pp. 152–157.
10. Thuan, N. D., Hai, V. D., Webster, J. G., and Nimunkar, A. J. (2011). 'A web-based electronic medical records and hospital information system for developing countries'. Journal of Health Informatics in Developing Countries (JHIDC). 5(1): 155–170.
11. Barni, M. et al. (ed.) Fractal image compression. Document and Image Compression. CRC Press, Boca Raton, FL, 2006, Vol. 968, pp. 168–169.
12. Welstead, S. T. Fractal and wavelet image compression techniques. SPIE Publication, Bellingham, Washington, 1999, pp. 155–156.
13. Huang, H. K. PACS and Imaging Informatics: Basic Principles and Applications. John Wiley and Sons, Inc., Publication, New York, 1999.
14. Hai, V. D., Thuan, N. D., Ngoc, P. P., Huy, Q. H., and Thanh, V. P. (2010). A design of renal dataflow control and patient record management system for renal department environment in Vietnam. In V. Van Toi, and T. Q. D. Khoa (eds.), The 3rd International Conference on the Development of Biomedical Engineering in Vietnam, IFMBE Proceedings, Vol. 27. Springer, Berlin, Heidelberg.
15. Hai, V. D., and Thuan, N. D. (2010). 'Design of Laboratory Information System for Healthcare in Vietnam BK-LIS'. The 3rd International Conference on Communications and Electronics (ICCE), pp. 110–114.
16. Hai, V. D., Pham, M. H., Lai, H. P. T., Dao, V. H., Nguyen, D. T., and Phan, D. H. (2017). 'Design of Software for Wireless Central Patient Monitoring System'. Proceedings of KICS-IEEE International Conference on Information and Communication with Samsung LTE&5G Special Workshop, pp. 214–217.

17. Hai, V. D. et al. (2020). Design of noninvasive hemodynamic monitoring equipment using impedance cardiography. In V. Van Toi, T. Le, H. Ngo, and T. H. Nguyen (eds.), 7th International Conference on the Development of Biomedical Engineering in Vietnam (BME7), BME 2018, IFMBE Proceedings, Vol. 69. Springer, Singapore.
18. MMED6000DP-S6 service manual, MMEDCHOICE (2008).
19. Ngoc, P. P., Hai, V. D., Bach, N. C., and Van, B. P. (2015). EEG signal analysis and artifact removal by wavelet transform. In V. Toi, and T. Lien Phuong (eds.), 5th International Conference on Biomedical Engineering in Vietnam, IFMBE Proceedings, Vol. 46. Springer, Cham.

11 Universal Health Database in India
Emergence, Feasibility and Multiplier Effects

Arindam Chakrabarty[1] and Uday Sankar Das[2]
[1]Department of Management, Rajiv Gandhi University
Doimukh, Arunachal Pradesh, India
[2]Department of Management and Humanities
National Institute of Technology Arunachal Pradesh
Yupia, Arunachal Pradesh, India

CONTENTS

11.1 Introduction 215
11.2 Health Information for Government Health Beneficiary 216
11.3 Universalisation of Vaccination 218
11.4 Compulsory Requirement of Vaccination for Visiting Abroad 219
    11.4.1 Review of Literature 220
    11.4.2 Objectives of the Study 225
11.5 Methodology 225
    11.5.1 Analysis I 226
    11.5.2 Analysis II 227
11.6 Conclusion 230
References 232

11.1 INTRODUCTION

The United Nations is committed to achieving 17 Sustainable Development Goals (SDGs) by 2030 [1]. Goal 3 categorically focuses on 'Good health and well-being'. In fact, it is one of the most challenging tasks that the global community has to accomplish. The health sector is considered one of the important pillars upon which a civilisation, society or nation rests. The situation of the basic health service is indeed vital, particularly in highly populous and developing countries like India. Policy makers, domain professionals and researchers have been advocating to develop and introduce a universal heath record system for every one of us.

India has been emerging as a transforming economy worldwide, but it has been observed that the budget allocation on health has been decreased by the government.

However, the private sector is taking keen interests in participating in this crucial mission. The 71st round of NSS has observed that the social consumption in the heath sector is extraordinarily high. It doesn't imply that the increasing pattern of expenditure on medical treatment alone is responsible for high contribution from households; rather, there are plenty of other factors like cost of transportation, expenditure incurred for stay of both the patient and escort, loss of worker hours during the period of treatment and even less productivity due to morbidity. Moreover, India has not achieved 100% medical insurance to all its citizens even though government has initiated a landmark scheme named 'Ayushman Bharat Yojana' in order to bring the lower strata of citizen under the ambit of health insurance in spite of regular advisory from the United Nations.

The dynamic changes in the environment and the growing influx of pollution it exacerbates the spread of certain complicated viral or bacterial infections that most particularly affect children and elderly population. The use of antibiotics has enormously increased. Now the domain experts are mostly concerned about the syndrome of resistance to antibiotics and lifesaving drugs. This is simply because we do not have comprehensive health database for each of us. As a result, we are about to face a tragic situation where, in spite of revolutionary contributions from science and technology for evolving with innovative molecules, the solutions cannot be administered or effective for human well-being due to drug resistance syndrome.

In this twenty-first century, we are enriched by a series of molecules, drugs, antibiotics and lifesaving medicines which can overcome various critical medical ailments if it is detected and treated in a timely manner. However, the patients are using excessive antibiotics and other generic drugs which compel them to become natural resistance. This minimises a catch-22 situation where we have the desired drugs that can treat the disease but it could not be either administered or it may not yield effective results in case of drug-resistant patients. This indicates the importance of instituting a universal health database, particularly in India. This chapter explores how the universal health database may potentially transform the health scenario in India.

11.2 HEALTH INFORMATION FOR GOVERNMENT HEALTH BENEFICIARY

Under the social security measures, pregnant women are covered with state-supported packages till the successful delivery of the infant and beyond. During this process, specific information of regular progression of each beneficiary is recorded and stored based on that the scheme benefits are extended at regular intervals as per the plan. This enables the development of a discrete health database for pregnant women. A health card is supported by a unique identification (UID) and barcode where the basic information about the subscriber is stored. Moreover, with the health card, the beneficiary has access to medical treatment from empanelled hospitals, and through this process, some health-related information pertaining to each hospital the patient visits would automatically be stored. This can be a small replica of the universal health database management system.

The government of India is trying extremely hard to bring every citizen of the country under the ambit of universal health coverage through various national-level

Universal Health Database in India

health insurance schemes, including Rashtriya Swasthya Bima Yojana (RSBY), Central Government Health Scheme (CGHS), Aam Aadmi Bima Yojana (AABY), Janashree Bima Yojana (JBY), Employment State Insurance Scheme (ESIS), Universal Health Insurance Scheme (UHIS) and Ayushman Bharat National Health Protection Scheme. Following is a description of each:

- Rashtriya Swasthya Bima Yojana (RSBY): This program was launched by the Government of India, Ministry of Labour and Employment to provide hospitalisation coverage for families below the poverty line with a registration of Rs. 30/- and a coverage of Rs. 30, 000/-, which extends up to five family members, including the head of the family, his spouse and three dependents. The government has fixed rates and packages for the hospitals for a large number of interventions. The insurer is selected by the state governments through competitive bidding.
- Central Government Health Scheme (CGHS): This aims at providing healthcare coverage for all employees and dependents of the Central Government of India through wellness centres.
- Aam Aadmi Bima Yojana (AABY): The scheme aims to provide social security to rural landless households covering the head of the family and one earning member. The annual premium is Rs. 200/- shared equally by the state and the Central Government. It provides Rs. 30,000/- for natural death, Rs. 75,000/- for accident or disability and Rs. 35,000/- for partial disability. The Central Government has created a separate fund maintained by LIC of India known as the Aam Admi Bima Yojana Premium Fund.
- Janashree Bima Yojana (JBY): The scheme replaces two social security schemes with one and aims to provide life insurance for people below the poverty line or marginally above the poverty line in the age group of 18 to 59 years identified under 45 low-income wage occupational groups listed on its official website.
- Employment State Insurance Scheme (ESIS): This is a multifaceted social security scheme to cover maximum possible scenarios of socioeconomic safety for the worker/labour population and their dependents, also taking care of temporary or permanent disability resulting from factory/ workplace accidents. It also aims to provide pensions to the dependent deceased's labour family in case of workplace death. It is applicable for factories employing 10 or more persons. The scheme also includes the hospitality industry, shops, theatres, motor-transport, news agencies, private nursing homes and educational intuitions employing 20 or more persons. The scheme has already been implemented in most of India except for Arunachal Pradesh, Manipur, Mizoram, Sikkim, Chandigarh and Delhi.
- Universal Health Insurance Scheme (UHIS): Universal Health Insurance Scheme is being implemented by the four general insurance companies of the public sector targeting poor families. It provides reimbursement Rs. 30,000/- for medical expenses, Rs. 25,000/- for death or accident and Rs. 50/- per day for 15 days to the decedent's family. The target group for this health insurance is below poverty line families and the premiums are

Rs. 200 for individual, Rs. 300 for a five-member family and Rs. 400 for a seven-member family [2].

- Ayushman Bharat National Health Protection Scheme: This is a pan-India health insurance scheme with an aim to cover beneficiaries of approximately 50 Cr., covering up to five lakh rupees per family per year for healthcare coverage. The coverage can be available anywhere in India from any of the empanelled hospitals for public/private sectors. It also encourages co-operative federalism and dilutes power to be controlled by the state government. Ayushman Bharat National Health Protection Mission Council (AB-NHPMC) is the apex body under the banner of the Ministry of Union Health and Family Welfare with the State Health Agency as its state-level partner. NITI Aayog is the partner agency for providing the IT platform for cashless and paperless operations. One of the possible impacts of this would be reduction in cash transactions for the poorest of the poor, larger families, rural population and urban worker families [3].

Although the government of India has been trying its best to bring everyone under the umbrella of health social security, there is a clear lack of unified command for understanding the needs, requirements and delivery of universal health coverage with a plethora of services being planned but without a needs assessment for individuals on a case-by-case basis. A single health database, on the other hand, could reduce the burden of multiple policy or insurance schemes.

11.3 UNIVERSALISATION OF VACCINATION

The World Health Organization (WHO) regularly maintains and updates a list of vaccine preventable diseases [4]. WHO also runs a campaign for world immunisation under the banner of World Immunization Week, celebrated in the last week of April, with an aim to promote and promulgate the use of vaccines for protecting people from all age groups through immunisation [5]. The Global Vaccine Action Plan (GVAP) is yet to be fulfilled despite several efforts by several member countries of WHO, and the continued existence of polio despite a vaccine being available for decades is a testament to this problem, as is its continuance 20 years past the its target deadline of total eradication. The general aim of the GVAP is categorised by two criteria: either the target is fulfilled or not. This approach doesn't recognise that there are several underlying problems with a varied range. The GVAP was developed through intense consultation amongst the member countries with a focused vision and global strategy for immunisation; however, it has limited capacity to persuade partner countries to focus on the goals in a top-down manner [6].

At the national level, India has been trying to achieve its Universal Immunization Programme in some form or another since the late 1970s. The program started with the Expanded Programme of Immunization (EPI) in 1978, followed by the Universal Immunization Programme (UIP) in the 1985, subsequently followed by Technology Mission On Immunization in 1986, Child Survival and Safe Motherhood (CSSM) in 1992, Reproductive Child Health (RCH 1) in 1997 and National Health Mission (NRHM) from 2005 to 2014 [7].

Universal Health Database in India

The Universal Immunization Programme initially listed seven vaccines: tetanus, pertussis, diphtheria, polio, measles, bacillus Calmette-Guerin (BCG) and hepatitis B, along with Japanese encephalitis in prevalent districts. The program has identified 0.4 million low-immunisation pockets with migratory population. The government conducts Special Immunization Weeks every year to help fix the low immunisation problem through monitoring by National Technical Advisory Group on Immunization, with a special focus on creating a list of beneficiaries [8].

The Universal Immunization Programme follows an 'National Immunization Schedule' and primarily targets infants ranging from birth to 12 months and providing BCG, hepatitis-B birth dose, OPV (oral poliovirus vaccine) birth dose, OPV 1, 2 and 3, IPV (inactivated polio vaccine), pentavalent 1, 2 and 3, rotavirus vaccine, measles first dose, vitamin A first dose, and children 16 months to 16 years ranging from DPT (diphtheria-pertussis-tetanus) first booster, OPV booster, measles second dose, vitamin A (second to ninth dose), DPT second booster and TT [9].

The Universal Immunization Programme now lists nine diseases for vaccination nationally and three diseases subnationally, along with the implementation of 'Mission Indradhanush', which aims at achieving 90% coverage for children. The govern is also focused on building capacity for stocking vaccines with the help of the National Cold Chain & Vaccine Management Resource Centre and the training of staff through the National Cold Chain Training Centre. This has led to the creation of Electronic Vaccine Intelligence Network or (eVIN) with the sole purpose of tracking the vaccine stock, including tracking the storage temperature on a pan-India basis and real-time monitoring of stocks with the help of National Cold Chain Management Information System to simultaneously monitor the equipment inventory, functionality and availability [10].

This fragmented approach towards administering, monitoring and recording vaccination influences the ideation of an integrated universal health database for India.

11.4 COMPULSORY REQUIREMENT OF VACCINATION FOR VISITING ABROAD

Human migration is an age-old phenomenon which includes travel for either business or leisure. This has helped spread of disease from one nation to other via trade routes. For instance, the Third Plague pandemic started in the Chinese southwestern region of Yunnan and slowly migrated to Hong Kong, Japan, Singapore, Rio de Janeiro, Honolulu, San Francisco, Sidney and Bombay, finally arriving at the doorstep of Europe in late 1896 [11].

Today humans have the ability to achieve long-distance travel with the help of cheap air transport, thereby becoming the transport agent of microbes, infections and biota across the planet. This has also created a problem of invasive species and destruction of habitat of native species, promoting the rise of pandemic diseases without any check mechanism. The Asian countries have become the new hot spot for tourism for western travellers due to low cost of stay and transport. The world population has reached the highest ever recorded over the centuries, with most of the people living in urban, cohabited and often cramped spaces. The proximity of animals and humans living side by side has also been reduced to minimal, creating an environment for easy spread of infectious diseases, along with the changing temperature and weather. Humans' ability to become carrier agents for infectious diseases through travel is evidenced by the spread of HIV, drug-resistant tuberculosis, vector-borne diseases like chikungunya and dengue emerging in the poorer third-world countries. Another prominent reason for travel is medical treatment in locations where the cost of healthcare is comparatively lower. Countries like Thailand, Singapore, Malaysia and India are some of the major medical tourism destinations; this has also attracted the attention of policy makers to look for linkages for possible spread of disease from one country to another [12].

India didn't have any health check requirement for outbound travellers until the COVID19 pandemic; however, any person arriving from any of the yellow fever infected country will be kept under quarantine for a period of six days if the person is not vaccinated. In case of novel diseases (e.g., COVID19) the quarantine period was followed as per global standards or WHO protocols [13].

However, WHO has a different perspective and lists several vaccinations for inbound travellers to India, namely adult diphtheria and tetanus vaccine, hepatitis A vaccine, hepatitis B vaccine, typhoid vaccine, oral polio vaccine, varicella vaccine, Japanese encephalitis vaccine, meningococcal vaccine, rabies vaccine and yellow fever vaccine [14].

WHO also maintains a list of territories and vaccination requirements for international travellers for maximum safety and precaution (Table 11.1).

11.4.1 REVIEW OF LITERATURE

The rise of database management systems (DBMS) has popularised the wide use of electronic health databases for heath research in the past decade. The databases are such that they regularly collect data from clinical practice, with a potential to contribute to academic research activities. The UK has pioneered in this field with the introduction of the General Practice Research Database (GPRD), which is the largest health database worldwide accessible to academicians globally. It came into being in 1987 as a databank known as Value Added Medical Products (VAMP). In 1994 it was transferred to the UK's health department and was formally named GPRD. It currently holds a record of around 5 million patients from 590 primary care practices in the UK.

The level of anonymity and automation in submission and extraction of medical records makes it a powerful analytical tool for health research. Such a public health database promotes scientific production in many ways, with reduced barriers for secondary analysis of data for researchers [17].

Health outcome information and utilisation is significantly valuable to healthcare researchers as well as policy makers for person-specific objectives. British Columbia has pioneered the field by building an administrative health database and linking all person registered with the healthcare system.

This linkage has helped achieve high success for matching person-specific registration, which has in turn helped to propose research projects that were otherwise unheard of. It has also helped address the issues related to security, ethics and confidentially with these valuable resources [18].

TABLE 11.1
Indicative List of Disease and Vaccination Required for Travel

Name of the Disease	Country of Travel	Source
Yellow Fever	Afghanistan, Albania, Algeria, American Samoa (USA), Andorra, Angola, Anguilla, Antigua And Barbuda, Argentina, Armenia, Austria, Azerbaijan, Bahamas, Bahrain, Bangladesh, Barbados, Belarus, Belgium, Belize, Benin, Bermuda, Bhutan, Bolivia, Bosnia And Herzegovina, Botswana, Brazil, British Virgin Islands, Brunei Darussalam, Bulgaria, Burkina Faso, Burundi, Cambodia, Cameroon, Canada, Cayman Islands, Central African Republic, Chad, Chile, China, China, Hong Kong SAR, China, Macao SAR, Christmas Island, Colombia, Comoros, Congo, Cook Islands, Costa Rica, Côte D'ivoire, Croatia, Cuba, Cyprus, Czech Republic, Democratic People's Republic Of Korea, Democratic Republic Of The Congo(Formerly Zaire), Denmark, Djibouti, Dominica, Dominican Republic, Ecuador, Egypt, El Salvador, Equatorial Guinea, Eritrea, Estonia, Ethiopia, Falkland Islands (Malvinas), Faroe Islands, Fiji, Finland, France, French Guiana, French Polynesia, Gabon, Gambia, Georgia, Germany, Ghana, Gibraltar, Greece, Greenland, Grenada, Guadeloupe (France), Guam (USA), Guatemala, Guinea, Guinea-Bissau, Guyana, Haiti, Honduras, Hungary, Iceland, India, Indonesia, Iran, Iraq, Ireland, Israel, Italy, Jamaica, Japan, Jordan, Kazakhstan, Kenya, Kiribati, Kuwait, Kyrgyzstan, Lao People's Democratic Republic, Latvia, Lebanon, Lesotho, Liberia, Libya, Liechtenstein, Lithuania, Luxembourg, Madagascar, Malawi, Malaysia, Maldives, Mali, Malta, Marshall Islands, Martinique (France), Mauritania, Mauritius, Mayotte (France), Mexico, Micronesia (Federated States Of), Monaco, Mongolia, Montenegro, Montserrat, Morocco, Mozambique, Myanmar (formerly Burma), Namibia, Nauru, Nepal, Netherlands, Netherlands Antilles (Bonaire, Curaçao, Saba, St Eustatius, St Martin), New Caledonia and Dependencies (France), New Zealand, Nicaragua, Niger, Nigeria, Niue, Northern Mariana Islands (USA), Norway, Oman, Pakistan, Palau, Panama, Papua New Guinea, Paraguay, Philippines, Pitcairn, Poland, Portugal, Puerto Rico (USA), Qatar, Republic of Korea, Republic of Moldova, Reunion (France), Romania, Russian Federation, Rwanda, Saint Barthelemy, Saint Helena, Saint Kitts and Nevis, Saint Lucia, Saint Martin (France), Saint Pierre And Miquelon (France), Saint Vincent and The Grenadines, Samoa, San Marino, Sao Tome and Principe, Saudi Arabia, Senegal, Serbia, Seychelles, Sierra Leone, Singapore, Slovakia, Slovenia, Solomon Islands, Somalia, South Sudan, Spain, Sri Lanka, Sudan, Suriname, Swaziland, Sweden, Switzerland, Syrian Arab Republic, Tajikistan, Thailand, The Former Yugoslav Republic of Macedonia, Timor-Leste, Togo, Tokelau, Tonga, Trinidad and Tobago, Tunisia, Turkey, Turkmenistan, Tuvalu, Uganda, Ukraine, United Arab Emirates, United Kingdom (with Channel Islands and Isle of Man), United Republic of Tanzania, United States of America, Uruguay, Uzbekistan, Vanuatu, Venezuela (Bolivarian Republic of), Viet Nam, Virgin Islands (USA), Wake Island (USA), Wallis And Futuna (France), Yemen, Zambia, Zimbabwe	[15]

(Continued)

222 Soft Computing Applications and Techniques in Healthcare

TABLE 11.1 *(Continued)*
Indicative List of Disease and Vaccination Required for Travel

Name of the Disease	Country of Travel	Source
Malaria	Afghanistan, Algeria, Angola, Argentina, Azerbaijan, Bangladesh, Belize, Benin, Bhutan, Bolivia, Botswana, Brazil, Brunei Darussalam, Burkina Faso, Burundi, Cabo Verde, Cambodia, Cameroon, Central African Republic, Chad, China, Colombia, Comoros, Congo, Cook Islands, Costa Rica, Côte d'Ivoire, Democratic People's Republic of Korea, Democratic Republic of the Congo (Formerly Zaire), Djibouti, Dominican Republic, Ecuador, Egypt, El Salvador, Equatorial Guinea Eritrea, Ethiopia, French Guiana, Gabon, Gambia, Georgia, Ghana, Gibraltar, Greece, Guatemala, Guinea, Guinea-Bissau, Guyana, Haiti, Honduras, India, Indonesia, Iran, Iraq, Kenya, Kyrgyzstan, Lao People's Democratic Republic, Liberia, Madagascar, Malawi, Malaysia, Mali, Mauritania, Mayotte (France), Mexico, Mozambique, Myanmar (formerly Burma), Namibia, Nepal, Nicaragua, Niger, Niue, Oman, Pakistan, Paraguay, Peru, Philippines, Republic of Korea, Russian Federation, Rwanda, Saint Lucia, Sierra Leone, Singapore, Somalia, South Africa, South Sudan, Sri Lanka, Sudan, Suriname, Swaziland, Syrian Arab Republic, Thailand, Uganda, Ukraine, United Republic Of Tanzania, Yemen, Zimbabwe.	[15]
Polio	Afghanistan, Pakistan, Nigeria, Ethiopia, Kenya, Somalia	[16]

Health professionals often identify access to public health database as problematic. Public health literature is often plagued by grey literature apart from the regular journals and articles. Databases such as Medline are inefficient and don't meet the needs and wants of public health workers or researchers in general. Results suggests that Global Health Database holds unique medical records in comparison to Medline, where a significant proportion of duplicity was found, along with test results which shows that Global Health Database as a complementary product [19].

In 1986 WHO initiated a global database on child growth and malnutrition with the sole purpose of collection and standardisation to disseminate anthropometric data of children. The database contains 846 surveys of 412 national surveys collected from 138 countries and 434 sub-national surveys of 155 countries. The database covers 99% in developing and 64% in developed countries of children under 5 years of age.

This rich data enables a comparison of nutritional value and the evaluation of nutritional needs and other government interventions to monitor child growth and create sociopolitical awareness of the issues relating to childcare. This database stands as an epitome of success for international collaboration for collection and hominisation of child growth data in the last 15 years [20].

The West Australia Linked Database links available administrative health data of 1.7 million residing in Australia with a diversity of six core data elements ranging from birth, midwives, cancer, morbidity of inpatients, in-outpatient mental health services and death. It is designed to be updated regularly and provide geo-codes. The system is a benchmark that upholds all international data for collection and collaboration [21].

Universal Health Database in India

Financial information of the health sector is essential for making informed decisions in order to reform the health sector. Healthcare financing planning should begin with a sound analysis of national expenditure on health, total cost, contributions from different sources and spending to be done for the allocation of the fund. OECD (Organisation for Economic Co-operation and Development) countries have shown significant progress in wise spending of these funds with the help of public private partnership. The United States, on the other hand, has formulated a comprehensive matrix named National Health Accounts which is an extension of the OECD methods. National Health Accounts portrays a holistic picture of the sector along with the key stakeholders, participants and the expenses incurred. This also helps in framing public policy, financing, monitoring, training and so on around health sector [22].

AADHAR Management System generates a 12-digit unique alphanumeric secure number for every Indian citizen, inclusive of infants and children. It aims to reduce the documentation process for both the individuals and government and make life easier. This system contains basic demographic information like name, sex, address, marital status, photo, ID mark, fingerprint biometric, iris (unique of human eye pattern) and signature. It aims to provide universal identity infrastructure usable for proof of identity (POI) and proof of address (POA). The steps in this direction began with the recommendation of 'Kargil Review Committee Report' of January 2000 that suggested immediate issuance of identity cards to prevent illegal immigrants from obtaining Indian ID cards.

With this recommendation and the subsequent report of ministers titled 'Reforming the National Security System' cited the need for a Multipurpose National Identity Card (MNIC) for segregating Indian Nationals and non-nations which resulted in the creation of National Register of Citizens in 2003 [23].

UID stands for 'unique identification', which is essentially a number assigned to every Indian citizen with the aim of 'one person, one number' which is verifiable via fingerprint biometric and is issued by UIDAI (the Unique Identification Authority of India). It aims to be non-duplicable. In 2009 the planning commission started the process under an executive order. It has two primary aims: first to create robust duplicity-free identity cards and second to made identity easy to verify and cost-effective. The UIDAI lists documents like PAN (permanent account number) and Voter ID as valid documents acceptable to prepare Aadhaar or UID number under its 'Handbook for Registrars'. Person who lacks all these documents may use 'introducers' to act as guarantee for issuing Aadhaar.

The then home minister referring to the security, resilience and seamlessness at a conference 'Intelligence Bureau Centenary Endowment, 2009' pointed to the use of 22 databases under National Intelligence Grid (NATGRID). That could be further extended for use by other intuitions like banks, insurance providers, telecoms and so on that require government-issued POI for authorisation and identification [24].

The penetration of information technology into the health sector has portrayed a promise of healthcare with security concern in e-health care domain and the role of UIDAI. Health is a primary issue for the vast geographical stretch of India along with the states and union territories. There is a severe lack of infrastructure and doctors especially in the rural areas, as shown by the statistic that 75% of doctors practice in urban areas and 23% in semi-urban areas.

There is also a lack of comprehensive health scheme coverage for all in the country even though India is a leader in telemedicine with over 380 centres. Challenges like poverty, illiteracy and the reach of Information and communications technology ICT has hindered the adoption of health in India. Premier medical institutions like AIIMS, Apollo, ISRO, DIT, CSIR, SGPI and CDAC are showing some promising developments in the adoption of e-health that includes 'Cloud Enabled E-health Center', 'Virtual Medical Kiosk', 'RFID-ITRM', 'Alcohol Web India', 'E-health Project at Punjab' and so on [25].

The Aadhaar scheme that uses biometrics as identification for Indian residents has a reach of over one billion people and an ever-expanding horizon that promises improved social inclusion, direct benefit transfer and reduced corruption in the Indian social architecture. Although it is too early to say if the system would act as a transformation agent, the initial voluntary enrollment was impressive. There is a serious concern about government surveillance generated by the fast expansion; however, it also promises to strengthen public service programs and address broader inequalities if implemented properly [26].

In July 2018, the NITI Aayog proposed National Health Stack (NHS) to enable IT services in the heath sector. That resulted in the 'National Digital Health Blueprint' (NDHB) report that was submitted to the Ministry of Health and Family Welfare (MoHFW) on 24 April 2019, which later was published in the public domain for comments and recommendations from stakeholders.

The National Health Policy 2017 promised the extensive use of digital technologies for adopting a service delivery mechanism and universal health coverage. Some of the resultants of the NHP 17 was Ayushman Bharat Yojana and Pradhan Mantri-Jan Arogya Yojna (PMJAY) and to set up 0.15 million 1.5 lakh Health and Wellness Centers for promoting primary healthcare. National Digital Health Eco-system (NDHE) can help overcome the hurdles of health informatics for developing economies. The creation of National Digital Health Blueprint (NDHB) rests on the shoulders of the key stakeholders like patient, care professional, care provider, health players, governing bodies, research bodies and pharmaceutical players. National Digital Health Blueprint imagines an entire ecosystem in the health sector to enable a varied range of services to all the stakeholders in a digital ecosphere [27].

In addition to this, MoHFW and various welfare and other research organisations conducted health surveys either in pan India or state-wide (Table 11.2).

This shows that government spends a lot of money to collect health-related information at regular intervals. However, the reports are prepared after the assessment period, not concurrently. There is not infrastructure or mechanism to get real-time data on the health sector. As a result, it is difficult to take preventive, presumptive steps so that the indicators of this sector can be brought under control. This envisages the introductory idea and concept paper on creating a Universal Health Database in India where conceptually every Indian's health-related information would be recorded and transferred to individual's database so that comprehensive treatment healthcare system can be adopted.

Universal Health Database in India

TABLE 11.2

Indicative List of Health-Related Surveys in India by NSSO

Round	Time	Area
7th Round	October 1953–March 1954	Information on morbidity.
11th to 13th Round	1956–1958	Information on morbidity.
17th Round	September 1961–July 1962	Examining alternative approaches of morbidity reporting.
28th Round	October 1973–June 1974	Morbidity became a part of the decennial surveys on social, Consumption.
35th Round	July 1980–June 1981	All-India Survey on Social Consumption.
42nd Round	July 1986–June 1987	All-India Survey on Social Consumption.
50th Round	July 1993–June 1994	All-India Survey on Social Consumption with health as a subject.
60th Round	January–June 2004	Morbidity and healthcare, including the problems of aged persons.
71st Round	January–June 2014	Social consumption with health, health, morbidity, along with education and ICT.

11.4.2 OBJECTIVES OF THE STUDY

1. To study and understand the concept of Universal Health Database.
2. To propose the modus operandi and Other Collateral benefits of Universal Health Database in Excelling Health System in India.

11.5 METHODOLOGY

The study is designed in such a fashion that it could converge to develop a dedicated Universal Health Database System for all Indian citizens. The study has been carried out using secondary information particularly various health statistics, NSSO data and other journal papers across the country and abroad. Based on these information inputs and insights, it is evident that the health sector is suffering from various angles. One of these stumbling blocks may be the lack of a Universal Health Database System for all its citizens. This chapter has attempted to devise a suitable model based on the relevant reliable and available secondary information. This study has been developed using secondary relevant information from the domain literature and reports.

11.5.1 ANALYSIS I

The report of 71st round working group of NSSO has clearly demonstrated that the household has to incur various costs to avail health services. It is equally applicable in case of government-run hospitals or health centres where the cost of treatment is minimal. The cost of transportation to visit hospitals or health centres, cost of lodging and boarding for the patient and his or her attendants and the cost of the loss of working days are exorbitantly impacting the social consumption of health. Moreover, it is observed that the patient does not have his or her previous health records, As a result, in every occasion the physician has to advise similar diagnostic tests. There is either no record or a fractured record of the variety and volume of medicine administered so far. All of this poor documentation creates confusion in the service delivery process.

Several studies have determined that the people of twenty-first century would not suffer because we do not have medicines or antibiotics to address life-threatening diseases; rather, it is the unrecorded and massive misuse or overuse of such drugs or antibiotics and analgesic molecules that make our body strictly resistant to those drugs in future. That means if the future treatment desires the use of such drugs, the patient body would not respond as depicted in Graph 11.1 [28, 29, 30, 31].

With the emerging complexities of life, growing pollution, work-related stress and lifestyle disorders, human societies are more susceptible to various forms of diseases. Some ailments like diabetes, hypertension, depression and so on have become epidemics that impact people in the twenty-first century and onward. These diseases have other correlations with many other functions of the body. There are impacts of family history on the likelihood of acquiring certain disease as a matter of inheritance or certain congenital ailments which are very specific to individual patients. There are creation allergens like specific-category drug allergies (sulphur drugs, etc.), lactose intolerance and so on which is specific to the patient. Many times, all of this patient-specific information or historical data is not communicated to the doctors due to lack of awareness, forgetfulness or while patient is in serious condition

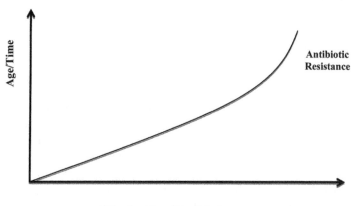

GRAPH 11.1 How overuse of antibiotics/drugs causes drug resistance over time.

Universal Health Database in India

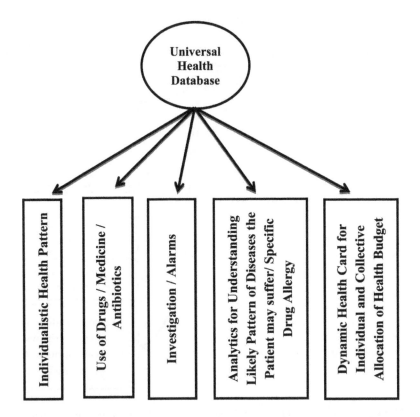

FIGURE 11.1 Indicative list of the nature of health-related information that would be captured in the UHD.

and family members do not know the information. All these are contributing factors to delay or derail medical treatment. This chapter has suggested that each individual should have a dedicated health database where all the health-related information is recorded instantly in real time basis so that during future medical treatment the doctor can access certain information protocol after acquiring appropriate authorisation.

An indicative list of activities that the proposed Universal Health Database should ideally comprise is shown in Figure 11.1.

11.5.2 ANALYSIS II

If adopted as a policy, the Universal Health Database should be made compulsory for all the stakeholders who would take part in service delivery mechanism in the health sector. All the portals of the service providers should compulsorily use a nationwide unique software package, and every service provider should prepare all the service offerings in the structured format using an individual UID (Aadhaar) number followed by appropriate authorisation from the patient or party. It may be operated on offline basis, and once it is connected to the internet, that health-related data would

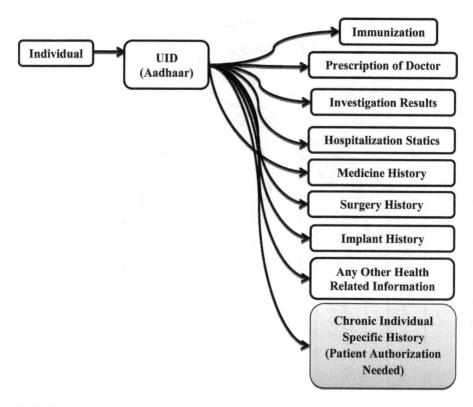

FIGURE 11.2 Indicative list of health-related information that would be captured in the UHD.

be transferred to the patient-specific Universal Health Database. The health-related data may include medicines prescribed by the doctor; however, in every case the patient consent is needed to access such a database and particularly to access very specific database information. Therefore, access remains at the discretion of the patient (Figure 11.2).

The Universal Health Database would be operationalised by devising a dedicated and comprehensive software package which would be installed in the portals of all health service providers. The indicative list of health service providers is prescribed in Figure 11.3. The service provider would record all the health-related inputs and transactions in respect of each UID, and when the software is connected to server via the internet, the UID-specific dataset would get transferred to the respective Universal Health Database.

The Universal Health Database would serve as a unique solution to many health-related issues. With the progression of life, every individual requires various forms of medical facilities starting from immunisation, doctors' visits, diagnostic investigations, lists of medicine procured and administered and many other health-related management issues that include lifestyle modifications. If all the health-related information of an individual is recorded and transferred to the database, future treatment would be immensely improved if past health information of

Universal Health Database in India

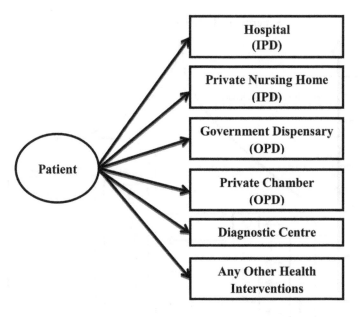

FIGURE 11.3 Indicative generic list of service providers where from the appropriate data would be collected and recorded in the UHD.

the patient can be adequately retrieved and processed. If the process of recoding continues during the entire life of the patient, it is expected that the patient would be greatly benefitted in the handling of critical care situations, problems associated with aging or and so on.

The universal health database would have two chambers. The first chamber would collect, collate, record and preserve the health-related information of any individual patient that could be retrieved instantly for further treatment or correlations of disease if it is authorised by the patient. The other chamber would act as logic units and may use the database for augmenting data analytics—namely, descriptive, diagnostic, predictive and prescriptive. Using the expectancy model, the data analytics would provide the most likely probabilistic line of disease or ailments that patients might suffer which would be closer to the actual incidences of disease suffered by the patients. That would make the clinical process simple but comprehensive and holistic. In Figure 11.4 jpg diagram, M_1, M_2, M_3,...Mn are the symbolic medical attentions or services taken by the patient, and in every case of medical attention, individualistic health data has been transferred to UHD either for future use or helping in the robust data analytics mechanism.

- Indicative Benefits:
    - Real-time recording of patient-specific health data.
    - Transparency in health-related transactions including financial data.
    - Selective access of the health database by doctors/physicians.
    - Robust mechanism of data transfer, storage, processing and retrieval.

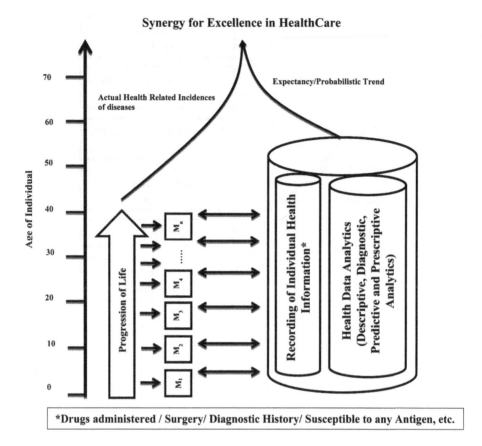

FIGURE 11.4 Interactive protocol of Universal Health Database: Expectancy modelling.

The individual UHD database can be connected with a national UHD (Figure 11.5) where without identifying the individual details the national UHD protocol would enable the retrieval of the appropriate dataset that would generate fundamental health indicators or case-specific status like immunisation, disease trends like malaria and cholera and so on in order to take preemptive, preventive and presumptive steps to control over epidemic or communicable diseases. This would help the state with its infrastructural preparedness to combat certain alarming health issues or diseases in a particular region.

11.6 CONCLUSION

This chapter has attempted to explore all these areas that would be immensely aided with the development and introduction of such comprehensive health database management systems. The comprehensive but individualistic health database system essentially helps physicians to derive any expectancy model for a person or community or even for a small region by means of exploring correlations, extrapolation of statistical datasets on various relevant attributes, and thereby it would generate significant likelihood of certain disease pattern and outcome. This would necessarily

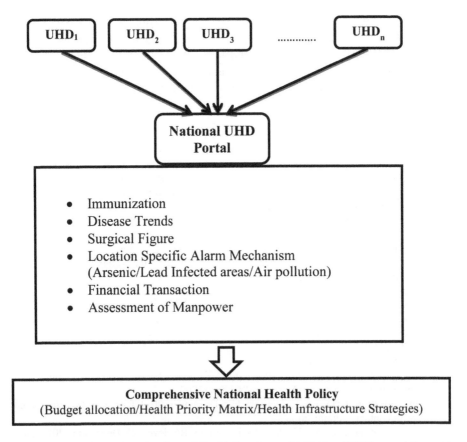

FIGURE 11.5 Linkages of Universal Health Database with National UHD portal for rapid policy reforms.

reduce the expenditure on healthcare and would be able to provide more proactive or curative measures rather than being restricted to curative health management. The Universal Electronic Medical Records System would allow the doctors to understand the various facets of patients like immunity standards, frequent infection patterns, identifications of allergens, history of previous ailments and drugs administered on a case-by-case basis. Universal Healthcare Database Management would immensely contribute towards research and development activities for medicinal, psychosocial or nutritional areas, among many others. The study has also assessed the extent of its feasibility, ease of access and ease of use particularly in an Indian context.

India is on the verge of becoming the fourth largest economy in the world and soon it may become the third largest after the United States and China. On the other hand, in terms of the Global Hunger Index, 2019 India ranks 102 out of 117 countries. This exhibits a dichotomous and diagonally opposite phenomenon in India where a tiny section of population enjoys the enormous wealth and large pool of population are denied even basic food and drinking water. In the doctrine of UN SDGs all the member countries are committed to achieve its 17-point agenda where poverty, hunger,

232 Soft Computing Applications and Techniques in Healthcare

well-being and education are the top priorities. At this critical juncture, the country desperately needs an individual, specific, real-time dynamic universal health database system which would enable planners and policy makers to work out contingency models along with the long-term plans. At the same time, the implementing agencies can get real-time feedback through its robust delivery mechanism.

REFERENCES

1. United Nations Department of Economic and Social Affairs. (n.d.). Retrieved 12 July, 2020, from https://sdgs.un.org/goals.
2. National Health Insurance Schemes. National Health Portal Of India. (n.d.). Retrieved 25 December, 2019, from https://www.nhp.gov.in/national-health-insurance-schemes_pg.
3. Ayushman Bharat - National Health Protection Mission. National Portal of India. (n.d.). Retrieved 25 December, 2019, from https://www.india.gov.in/spotlight/ayushman-bharat-national-health-protection-mission.
4. Vaccine-preventable diseases and vaccines 6.1 General considerations. (n.d.). Retrieved 25 December, 2019 from http://www.who.int/injection_safety/global-campaign/injection-safety_guidline.pdf.
5. World Immunization Week 2019. (n.d.). Retrieved 25 December, 2019, from https://www.who.int/campaigns/world-immunization-week/world-immunization-week-2019.
6. Strategic Advisory Group of Experts on Immunization. Review and Lessons Learned. (n.d.). Retrieved 25 December, 2019 from https://apps.who.int/iris/handle/10665/329097
7. Universal Immunization Program. (n.d.).
8. Universal Immunization Programme. (n.d.). Retrieved 25 December, 2019, from https://pib.gov.in/newsite/PrintRelease.aspx?relid=112060.
9. Immunisation. NHM. (n.d.). Retrieved 25 December, 2019, from http://www.nrhmhp.gov.in/content/immunisation.
10. Immunization: National Health Mission. (n.d.). Retrieved 25 December, 2019, from https://nhm.gov.in/index1.php?lang=1&level=2&sublinkid=824&lid=220.
11. Bramanti, B., Dean, K. R., Walløe, L., & Stenseth, N. C. (2019). The third plague pandemic in Europe. In Proceedings of the Royal Society B: Biological Sciences, 286(1901). https://doi.org/10.1098/rspb.2018.2429.
12. Travel, Conflict, Trade, and Disease - Infectious Disease Movement in a Borderless World - NCBI Bookshelf. (n.d.). Retrieved 26 December, 2019, from https://www.ncbi.nlm.nih.gov/books/NBK45724/.
13. Bureau of Immigration, G. of I., & Immigration Visa, F. R. and T. (n.d.). Health Regulation. Retrieved 26 December, 2019, from https://boi.gov.in/content/health-regulation.
14. Verma, R., Khanna, P., & Chawla, S. (2015). Recommended vaccines for international travelers to India. *Human Vaccines and Immunotherapeutics, 11*(10), 2455–2457. https://doi.org/10.4161/hv.29443.
15. List of countries, territories and areas. Vaccination requirements and recommendations for international travellers, including yellow fever and malaria. (n.d.). Retrieved from http://www.who.int/ith.
16. Ministry of Health and Family welfare, G. of I. (n.d.). Requirement of Polio Vaccination for International Travellers between India and Polio Infected Countries. Retrieved 26 December, 2019, from https://mohfw.gov.in/sites/default/files/08285260748Requirement.pdf.
17. Chen, Y. C., Wu, J. C., Haschler, I., Majeed, A., Chen, T. J., & Wetter, T. (2011). Academic impact of a public electronic health database: bibliometric analysis of studies using the general practice research database. *PloS one, 6*(6), e21404.

18. Chamberlayne, R., Green, B., Barer, M. L., Hertzman, C., Lawrence, W. J., & Sheps, S. B. (1998). Creating a population-based linked health database: a new resource for health services research. *Canadian Journal of Public Health, 89*(4), 270–273.
19. Aalai, E., Gleghorn, C., Webb, A., & Glover, S. W. (2009). Accessing public health information: a preliminary comparison of CABI's Global Health database and Medline. *Health Information & Libraries Journal, 26*(1), 56–62.
20. De Onis, M., & Blössner, M. (2003). The World Health Organization global database on child growth and malnutrition: methodology and applications. *International Journal of Epidemiology, 32*(4), 518–526.
21. Holman, C. D. A. J., Bass, A. J., Rouse, I. L., & Hobbs, M. S. (1999). Population-based linkage of health records in Western Australia: development of a health services research linked database. *Australian and New Zealand Journal of Public Health, 23*(5), 453–459.
22. Berman, P. A. (1997). National health accounts in developing countries: appropriate methods and recent applications. *Health Economics, 6*(1), 11–30.
23. Siddiqui, A. U., & Singh, M. H. K. (2015). "Aadhar" Management System. *IITM Journal of Management and IT, 6*(1), 40–43.
24. Chacko, S., & Khanduri, P. UID for Dummies. New Delhi, India: Jawaharlal Nehru University, 2011.
25. Srivastava, S., Agarwal, N., Pant, M., & Abraham, A. (December, 2013). A secured model for Indian e-health system. In 2013 9th International Conference on Information Assurance and Security (IAS), pp. 96–101, IEEE.
26. Bhatia, A., & Bhabha, J. (2017). India's Aadhaar scheme and the promise of inclusive social protection. *Oxford Development Studies, 45*(1), 64–79.
27. National Digital Health Blueprint (NDHB), Government of India, Ministry of Health & Family Welfare, eHealth Section, from https://www.nhp.gov.in/NHPfiles/National_Digital_Health_Blueprint_Report_comments_invited.pdf.
28. Pallasch, T. J. (2003). Antibiotic resistance, *Dental Clinics of North America, 47*(4), pp. 623–639. https://doi.org/10.1016/S0011-8532(03)00039-9.
29. Ventola, C. L. (2015). The antibiotic resistance crisis: part 1: causes and threats, *Pharmacy and therapeutics, 40*(4), 277.
30. Martens, E., & Demain, A. L. (2017). The antibiotic resistance crisis, with a focus on the United States, *The Journal of Antibiotics, 70*(5), 520–526.
31. Centers for Disease Control and Prevention. (2019). About Antibiotic Resistance, *Antibiotic/Antimicrobial Resistance,* https://www.cdc.gov/drugresistance/about.html.

12 Cluster Analysis of Breast Cancer Data Using Modified BP-RBFN

Viswanathan Sangeetha[1], Jayavel Preethi[2], Raghunathan Krishankumar[1], Kattur S. Ravichandran[1] and Ramachandran Manikandan[1]

[1]School of Computing, SASTRA Deemed University, Thanjavur, Tamil Nadu, India
[2]Department of Computer Science
Anna University Regional Center
Coimbatore, Tamil Nadu, India

CONTENTS

12.1 Introduction 235
12.2 Related Works 236
12.3 Proposed Back Propagation-Based Radial Basis Function Network 238
    12.3.1 Dataset Description 238
    12.3.2 Problem Description 238
    12.3.3 Feature Subset Selection 239
    12.3.4 Performing Clustering Using BP-RBFN 243
        12.3.4.1 Modified BP-RBFN 244
        12.3.4.2 Backpropagation Learning & Training Algorithms 245
        12.3.4.3 Working of BP-RBFN 246
    12.3.5 Cluster Validation 247
12.4 Experimental Results and Analysis 247
    12.4.1 Experimental Setup 247
    12.4.2 Performance Comparison 247
    12.4.3 Discussions 251
12.5 Conclusion 251
Acknowledgement 253
References 253

12.1 INTRODUCTION

According to the World Health Organization, breast cancer has become a common cancer type among women. It has been reported that in 2018 627,000 women died from breast cancer, which is approximately 15% of all cancer deaths among women. There are different modes of treatment for cancer, such as surgery, chemotherapy and radiation therapy [1]. Though there are treatments in the present scenario, early diagnosis will help toward a successful prognosis in a cancer case [2]. The treatment modalities differ for each stage of cancer. Availability of treatment and early diagnosis have increased the survival rate of cancer patients. Data-driven models for analysing the survivability of patients will help cancer management across the globe. Analysing the patterns that emerge from a group of data will undoubtedly help in knowledge discovery related to the diagnosis. Clustering is one such technique that divides the data into meaningful groups and aids in analysing the patterns that come out of it. Given a set of data points, clustering groups those with similar kinds of data points. The ultimate aim of clustering is to group unlabelled data innately. Clustering falls under unsupervised learning and helps researchers gain valuable insights from the given data.

This chapter proposes a data-driven model for breast cancer data analysis. Initially, a dataset from the UCI machine learning repository, Wisconsin Breast Cancer (Diagnostic) Dataset [3], is passed to the feature selection module. The dataset may contain certain features that will not contribute to the end analysis. In such cases, the dataset undergoes feature selection to eliminate the merely essential features. There are many feature selection methods in the literature. In clustering, they are classified as filter methods [4–7], wrapper methods [8, 9], hybrid methods [10–12] and embedded methods [13]. Each technique differs from the other with regard to its search strategy and selection criteria. Filter methods do not have criteria for selection. They are heuristic in nature and problem-specific. Euclidean distance [7] is used for its searching strategy. Wrapper methods use data mining techniques-based induction algorithms for their selection. Hybrid methods combine both wrapper and filter methods. They wrap the data using induction techniques and filter them using searching. Finally, once the essential features are selected, they are passed on to a clustering model. There are many clustering algorithms in the literature. In this chapter, a back propagation-based radial basis function neural network is used to group the unlabelled data based on the selected important features. The network is trained using six training algorithms to determine the best training algorithm. Clustering results obtained using all the training algorithms are analysed using performance measures like MSE index, the best performance in terms of time taken, best epoch, regression fit. The performance of BP-RBF under different training algorithms is analysed used the regression plot and MSE plot.

12.2 RELATED WORKS

Many works were found in the literature about cluster analysis of medical data. Dubey et al. [14] analysed data from the Wisconsin Breast Cancer Dataset using the k-means algorithm with computation measures like distance, centroid, epoch, split method, attribute and iteration to identify the correct metric that gives higher accuracy when used for diagnosis. An analytical model for predicting the survival rate of breast cancer patients was developed by Shukla et al. [15] by creating cohort clusters using unsupervised data mining methods. Using methods like SOM (Self-Organising Map) and DBSCAN (Density-Based Spatial Clustering of Applications with Noise), cohort clusters were formed. An MLP (Machine Learning Platform) is trained using these cohort clusters to analyse and predict the survivability of patients. Montazeri et al. [16] developed a rule-based classification method for predicting different types of breast cancer survival. Patterns that are found among datasets are used to determine the outcome of a disease. Popular machine learning techniques like Naive Bayes, nearest neighbour, support vector machine (SVM) with 10-fold cross validation technique were utilised along with the proposed model for prediction. Vivek Kumar et al. [17] analysed different classification techniques with conventional machine learning techniques. From the analysis, a prediction model with the best machine learning technique may be built for accurate prediction. Ahmed Idbal Pritom et al. [18] proposed an efficient feature selection and classification method for breast cancer prediction. Ranker method-based feature selection was used to rank the features. Using the N best-selected features, naive Bayes, C4.5 decision tree, and SVM were used to proceed with classification. Qiqige Wuniri et al. [19] proposed a generic-driven Bayesian classifier for breast cancer classification. The method was proposed to handle both the discrete and continuous types of features. Genetic algorithm (GA) was used to determine optimal feature subsets with a good area under curve metric. A GA-based feature selection was proposed by Fadzil Ahmad et al. [20] for diagnosing breast cancer. The GA-based selected features were classified using a parameter optimised artificial neural network (ANN) for further diagnosis. The network was trained with different training algorithms for performance analysis. Sirage Zenyu et al. [21] proposed a method for prediction of chronic kidney disease using ensemble methods. Info gain subset evaluator and wrapper subset evaluator were used to perform feature selection. Diagnosis was performed using k-nearest neighbour, J48, ANN, naïve Bayes and SVM. A breast cancer prediction system was proposed by Alickovic et al. [22] using GA- based feature selection and random forest-based classifier. An accuracy of 98% was obtained after feature selection. F-score method-based feature selection for breast cancer diagnosis was proposed by Akay et al. [23]. SVM was used for further classification of breast cancer data. In order to balance the training data, an undersampling technique was proposed by Liu et al. [24] for breast cancer prediction. Prediction was performed using decision tree at greater accuracy. particle swarm optimisation-based diagnosis system was designed by Sheikhpor et al. [25] for breast cancer prediction. A nonparametric kernel density estimation was integrated with particle swarm optimisation to achieve greater results. Kahkashan Kouser et al. [26] proposed a genetic algorithm-based feature selection method for clustering high-dimensional data. Initially the traditional k-means clustering algorithm was applied on the full dimension space and compared with the GA-based method. Zhiwen yu et al. [27] proposed a genetic based k-means algorithm for feature selection. The parameters of the algorithm were initialised using a weighting junction for better performance.

Many algorithms exist for feature selection, and all these processes pick the best features out of the existing feature set. Existing methods are broadly classified as *filter and wrapper* methods. The third class of method called the *hybrid* method combines both filter and wrapper method to get the best results. In the

238 Soft Computing Applications and Techniques in Healthcare

TABLE 12.1
Brief Survey of Popular Feature Selection Methods

Ref #	Method	Method Category	Data	Year
[28]	Modified bat algorithm	Filter	Wisconsin Diagnosis Breast Cancer Dataset	2017
[29]	GA-based feature selection	Wrapper	Wisconsin Breast Cancer Dataset, Wisconsin Diagnosis Breast Cancer Dataset, Wisconsin Prognosis Breast Cancer Dataset	2016
[30]	Graph-based feature selection	Wrapper	Wisconsin Diagnosis Breast Cancer Dataset	2018
[31]	Genetic programming	Wrapper	Wisconsin Breast Cancer Dataset	2019
[32]	Wrapper-based gene selection with Markov blanket	Wrapper	Colon, SRBCT, Leukemia, DLBCL, Bladder, Prostrate, Tox, Blastomi	2017

literature, many hybrid methods are proposed. Table 12.1 lists out some of the popular methods.

12.3 PROPOSED BACK PROPAGATION-BASED RADIAL BASIS FUNCTION NETWORK

12.3.1 Dataset Description

The Wisconsin Breast Cancer Dataset (diagnostic) is used in this work. The features are obtained from the digitised breast image of the FNA (fine needle aspiration) of a breast mass. The dataset has 569 instances and 32 features. Features include characteristics of the nuclei like a radius from canter to the perimeter, texture, perimeter, area, smoothness, compactness, concavity, number of concave points, symmetry and fractal dimension of each nucleus. The attribute and their domains are as given in Table 12.2.

12.3.2 Problem Description

The architecture of the proposed system is given in Figure 12.1. Wisconsin Breast Cancer (Diagnostic) Dataset from the UCI Machine Learning repository

Cluster Analysis of Breast Cancer Data Using Modified BP-RBFN

TABLE 12.2
Dataset

#	Feature	Domain
1	Sample Number	Id number
2	Thickness of clump	1–10
3	Cell size uniformity	1–10
4	Cell shape uniformity	1–10
5	Marginal adhesion	1–10
6	Size of single epithelial cell	1–10
7	Bare Nuclei	1–10
8	Bland chromatin	1–10
9	Normal nucleoli	1–10
10	Mitoses	1–10

has been used for simulations. The input data has as many as 569 instances and 32 features. The first and second, being ids, are ignored. Processing with the remaining 30 features will yield an entire cluster. Of the 30 elements, 10 features are selected using a genetic algorithm. The features are passed on to feed forward back propagation neural networks for training and testing. The training dataset is fed as input, which trains the proposed model, and later, the test dataset containing nine features is passed for testing the efficacy of the model. The network is trained using six training algorithms, and its behaviour is evaluated. RBF function is applied to the hidden layers to increase the relative change in weight and efficiency. The results are analysed through MSE and R-value to determine the fitting of output to target. Of all the methods, training using resilient back propagation was found the best in data fitting, with an average R-value of 0.95 on 25,50,100 epochs in 0.02 seconds with four hidden layers.

12.3.3 Feature Subset Selection

The accuracy of diagnosis depends on the relevancy of features. Relevant and irredundant features will help in accurate diagnosis. Eliminating the curse of dimensionality is one of the important steps in computer-aided diagnosis systems. The fineness of cluster depends on the features that are used for performing clustering. For a dataset S, 2^s number of subsets are possible. Processing with the optimal number of features is essential for the correct diagnosis. Once the dataset is preprocessed, the essential features are selected from the total available features. There are many techniques in the existing literature for feature selection. One of the prominent techniques that is used for feature selection is genetic algorithm [33, 34]. Genetic algorithm is an evolutionary optimisation-based approach that mimics the natural evolution process. Figure 12.2 shows the overall working strategy of genetic algorithm. It is a search heuristic that follows the natural evolution

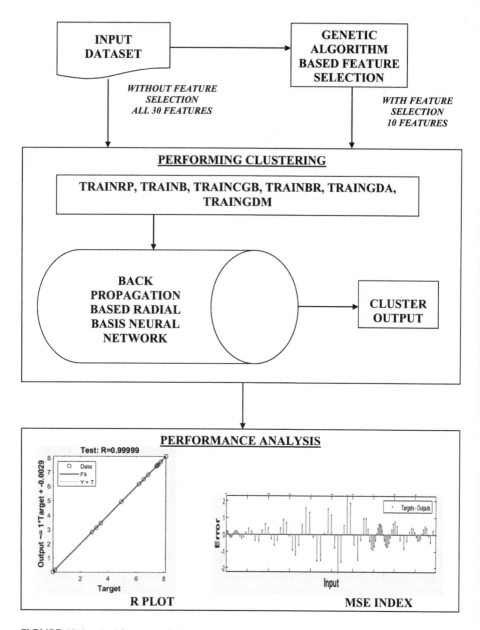

FIGURE 12.1 Architecture of the proposed system.

process of selecting the best individuals from the current population for further reproduction. Initially, the fittest individuals are selected from the population. They produce offspring that inherit the characteristics of their parents. They will have better fitness than their parents and thus survive at further

Cluster Analysis of Breast Cancer Data Using Modified BP-RBFN

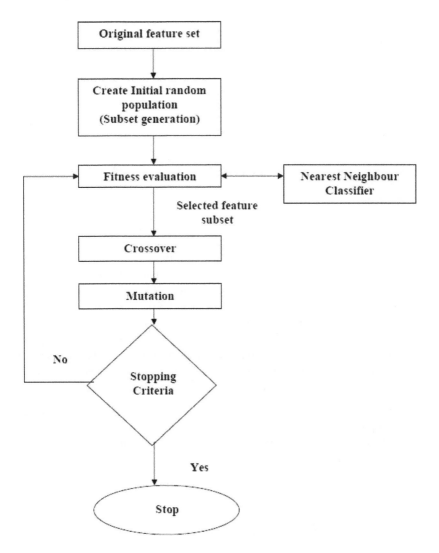

FIGURE 12.2 Architecture of genetic algorithm.

reproductions. This notion is applied for searching the best feature from the available list of features. The process of genetic algorithm has five stages: initial population creation, fitness function evaluation, selection, crossover and mutation. Once the initial population is created, the fitness of all individuals is evaluated. The best individuals are selected using roulette wheel selection. Using single-point crossover, the individuals are exchanged to result in new individuals. This routine is stopped when all the parent characters have been inherited by the offspring. Mutation is done on the individuals by exchanging their values internally.

The pseudo code for genetic algorithm is as follows:

> *begin*
> *initialise parameters*
> *generate random initial population*
> **while** *(not termination)*
> *determine the fitness of each individual*
> *select the individuals that satisfy the fitness evaluation criteria*
> *execute crossover*
> *execute mutation*
> *generate new population – selected individuals after crossover and mutation*
> **end while**
> **end begin**

Since genetic algorithm requires the solution to be represented in as strings for each chromosome, each solution is represented in the form of strings. The features are coded into a string form using binary coding scheme. Each chromosome is a possible solution represented as a string of length n bits. An ith bit represents ith feature of the feature subset. The feature is included if the value of bit in the string is 1. It is discarded if the value is 0. The fitness of each solution is determined using a nearest neighbours classifier (NNC) [33]. Each chromosome in the entire population represents an NNC configuration that shall undergo further evaluation. Searching through the pool of chromosomes is performed using genetic operators; crossover and mutation. Following is the procedure of GA-based feature selection:

1. From the total number of features, random feature subsets are created as initial population.
2. The features are encoded into a string to form a chromosome. The value of bit is 1 if features is included else, 0.

   Features

   $$F_1, F_2, F_3, F_4, F_5, \ldots\ldots\ldots\ldots\ldots F_{n-1}, F_n$$

   Encoded string

   $$1, 0, 1, 0, 1\ldots\ldots\ldots\ldots\ldots 0, 1$$
   $\downarrow$

   Final features as
   Chromosome

   $$F_1, F_3, F_5, \ldots\ldots\ldots\ldots\ldots, F_n$$

3. Fitness function: Once the feature subset is created, the strings are evaluated for its fitness value. The fitness is determined using NNC. Features having higher classification error are eliminated. and those with least classification error is passed on for further genetic operations. For a chromosome q, with the fitness function f, the fitness value can be stated by Equation (12.1).

$$f(q) = NNC(X, Tr_X, Tt_X) \qquad (12.1)$$

Cluster Analysis of Breast Cancer Data Using Modified BP-RBFN **243**

where NNC is the nearest neighbour classifier. X is the subset of entire feature space that has 1 in the encoded form, Tr_X is the set of training samples and Tt_X is the set of test samples.

NNC [33] can be given by Equation (12.2) as,

$$NNC(X, Tr_X, Tt_X) = \sum_{k=1}^{t} \sum_{i=1}^{N} f(Tt_i^k) * 100 / N_{Tt} \tag{12.2}$$

$$f(Tt_i^k) = \begin{cases} 1 \ if \ i = l^* \\ 0 \ otherwise \end{cases} \tag{12.3}$$

4. Fitness evaluation: Those feature subsets with lower classification error are selected for the genetic operations.
5. Crossover and mutation: Single-point crossover is applied on the selected feature subsets. After crossover between two offspring, bit string mutation is applied on the crossed offspring.
6. Generating new population: A new population is created with the new off-spring for the next round of genetic algorithm.
7. The process is stopped until the maximum number of iterations are met.
  Genetic algorithm is meta-heuristic in nature and problem independent. Thus it allows for the algorithm to be adapted to any problem.

12.3.4 Performing Clustering Using BP-RBFN

The selected features are passed on the next phase: clustering [35, 36]. Clustering is done to group the similar kind of data. Based on the selected features, the data points are grouped. Consider C_1 and C_2 as two clusters. A modified BPRBFN is used in this work to perform clustering. Let d_1 be the distance between X_1 and C_1, d_2 be the distance between X_1 and C_2. A point X_1 will belong to C_1 if $d_1 < d$.

A feedforward neural network (FNN) consists of a series of layers of interconnected processing unit neurons that pass information to each other via interconnections. The hidden layer neurons wait for the input from the input layer. Information is sent to the output layer once it is processed by the hidden layer. The transfer function is passed on to the hidden layers to map the input with the output. The number of hidden layers is also a factor for neural network usage. Figure 12.3 shows the architecture of a feed-forward neural network. Determining the number of neurons in the hidden layer is critical for a neural network design. It may result either in overfitting or underfitting. There are many methods in the literature for determining the number of neurons in the hidden layer. Try and error method [37] otherwise called Train and test method has been used in this work to determine the number of neurons.

The learning rate is also a critical factor in neural network architecture. A larger learning rate results in faster weight changes; thus, the network is trained faster. But too high a learning rate results in the objective function to diverge, thus resulting in no learning at all. Through a series of simulations, the value of the learning rate is taken in MATLAB®. It is considered as 0.25 here.

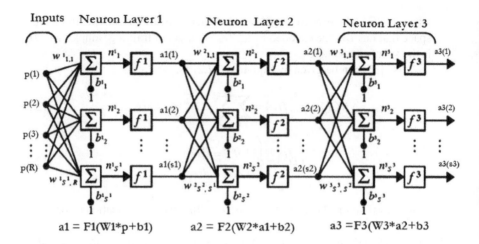

FIGURE 12.3 A feedforward neural network.

The weights from the input layer is calculated by hidden layer and transferred to the output layer. The amplitude of the output is limited by the activation function applied to the hidden layer.

The working of an FFN is explained in the following manner.

The basic processing unit of a neural network are the neurons.

1. The neuron x_{ij} gets the input and passes to the network.
2. On the interconnections, the inputs are added using Equation (12.4) and sent to hidden layer. The addition is performed using a linear combiner.

$$u = \sum_{i=1}^{n} x_i w_{ij} + b \tag{12.4}$$

3. On this result, the activation function is performed and passed on to the output layer. The activation function will help in reducing the amplitude so as to map the result to the output.

$$y = \varnothing(u + b) \tag{12.5}$$

12.3.4.1 Modified BP-RBFN

A radial basis function network is a variant of feedforward neural network with a radial basis function as a transfer function [38]. The selected features are passed on as input to the input layers. The transfer function of RBFN is as given in Equation 12.6.

$$f(x) = \varphi(\|x - \mu\|) \tag{12.6}$$

In this equation, x denotes the input vector, μ denotes the centre of radial basis functions, and $\|\ \|$ denotes the Euclidian distance.

Cluster Analysis of Breast Cancer Data Using Modified BP-RBFN 245

Typical radial basis function can be Gaussian and logistic function as given in Equations (12.7) and (12.8)

$$f(x) = \varphi(\|x - \mu\|) = \exp\left[\frac{-\|x - \mu_i\|^2}{2\sigma_i^2}\right] \qquad (12.7)$$

$$f(x) = \varphi(\|x - \mu\|) = \frac{1}{1 + exp\left[\frac{-\|x - \mu_i\|^2}{\sigma_i^2}\right]} \qquad (12.8)$$

where σ is the amplitude that determines the maximum distance between two nodes.

One of the significant challenges in designing an RBF is the determination of spread constant. When spread constant is too high, there will be overfitting of the network. When it is too low, there will be an underfitting of the network. To resolve this issue, this modified RBF has PSO to determine the spread consistently. This hybrid RBF is coupled with BNN for further training.

12.3.4.2 Backpropagation Learning & Training Algorithms

Training is usually done to perform an operation by adjusting the weights between neurons. A training example consists of a unique signal that shows either maximising or minimising to match the output against the target. Iterations continue until the goal is matched. Here, six training algorithms [39, 40] are used and compared. They are categorised as batch training and gradient descent training. Batch training proceeds by making the weight and bias changes in the network with the entire set of input matrices. Gradient descent training is an extension of the Widrow-Hoff algorithm in the backpropagation algorithm. It attempts to minimise the cost function of network learning equal to the MSE of the actual and target values. In this case, it tries to find the gradient of the error in the cost function. The cost function in backpropagation learning is as given in Equation (12.9):

$$E = \sum_{n=1}^{p} E(n) \qquad (12.9)$$

$$\text{where, } E(n) = \frac{1}{2}\sum_{i=1}^{u}(d_i(n) - o_i(n))^2$$

Gradient of cost function, $\nabla E = \dfrac{\partial E}{\partial W_1}, \dfrac{\partial E}{\partial W_2}, \ldots \ldots . \dfrac{\partial E}{\partial W_L}$, L = number of weights.

The weights in each iteration can be updated according to the weight update rule using Equation (12.10):

$$W_{ij}(n+1) = W_{ij}(n) - \eta\frac{\partial E(n)}{\partial W_{ij}} \qquad (12.10)$$

The main objective of the training is to minimise the total error function, which is the difference between the actual and target pattern.

The six training algorithms are as follows:

1. Trainrp: Training using resilient backpropagation
2. Trainb: Batch training
3. Traincgb: Training using conjugate gradient backpropagation with Powell-Beale restarts
4. Traingda: Training using gradient descent with adaptive learning rate backpropagation
5. Traingdm: Training using gradient descent with momentum backpropagation
6. Trainbr: Training using Bayesian regularisation backpropagation

12.3.4.3 Working of BP-RBFN

Figure 12.4 shows a sample working of BP-RBFN. Each neuron is excited only once. The linear function of the weighted kernel function combination ε is the membrane potential of the neuron. The internal state associated with each neuron is calculated using Equation (12.11):

$$u_{ij} = b + \sum_{i=1}^{n} x_i w_{ij} \varepsilon(t) \tag{12.11}$$

Let k be the delay associated with each neuron. Then by Equation (12.12),

$$u_{ij} = b + \sum_{i=1}^{n} x_i w_{ij} \varepsilon(t) d^k \tag{12.12}$$

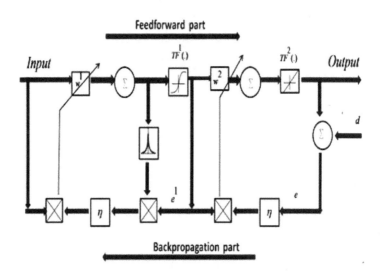

FIGURE 12.4 Working of BP-RBFN.

Cluster Analysis of Breast Cancer Data Using Modified BP-RBFN 247

12.3.5 CLUSTER VALIDATION

The final results of the algorithms are evaluated using three measures: mean squared error (MSE), R-value and R fit. The performance of the feature selection process is also checked by using the above-mentioned measures. The regression analysis compares the actual output values with the target ones. The resultant correlation coefficient R is returned. It also returns the best-fit equation of R. R has the value from [0.0, 1.0]. Higher the value of R, the greater the accuracy in the response of the network. Mean squared error returns the difference between actual and expected values. It is calculated using Equation (12.13):

$$\text{MSE} = \frac{\sum_{j=1}^{N}(y_j - o_j)^2}{N} \tag{12.13}$$

The error is computed by repeating the experiment for different numbers of iterations.

12.4 EXPERIMENTAL RESULTS AND ANALYSIS

12.4.1 EXPERIMENTAL SETUP

All the stages in experiments are performed in MATLAB® 2017b [41] in a Windows 10 with a core i7 processor and 12GB RAM. The training algorithms were coded in MATLAB® and experimented using the ANN toolbox. The preprocessing of the dataset is carried using the MATLAB® GUI. The script of GA has been coded in MATLAB® and tested. The training algorithms tested under different epochs. Table 12.3 shows the parameter values used for simulation.

12.4.2 PERFORMANCE COMPARISON

The training parameters [31] for each training function are listed in Table 12.3. The time consumed for the processing of each algorithm is as given in Table 12.4 under the best performance. But comparatively, resilient backpropagation was seen to be a bit faster and exhibit an accurate fit. Then, conjugate algorithms were seen to be second fastest with its minimal step size. However, for small networks, trainbr and batch training would be appropriate because they could converge faster with more minor updates. Resilient backpropagation converges at a quicker rate with larger weights. Graphs 12.1 and 12.2 show the plots of MSE and regression analysis. Graph 12.1 shows the mean squared error of the performance comparison. It can be found that the Bayesian regularisation training algorithm shows the minimum error of a different number of iterations. But the best fit and regression analysis of Bayesian regularisation is not as good as resilient backpropagation. Also, the mean squared error of resilient backpropagation is the second best on average under all number of iterations. Graph 12.2 shows the regression fit graph of the performance comparison. It can be seen that resilient back-propagation shows the best fit to the maximum at all number of iterations.

The final cluster diagrams of the network are shown in Figure 12.5.

Figure 12.5 (a–f) shows the cluster distribution of all training algorithms on BPRBFN. The cluster distribution of trainrp (Figure 12.5f) is very distinct when compared with other algorithms.

TABLE 12.3
Training Parameters of Training Functions

Parameters	BPRBFN-B	BPRBFN-CGB	BPRBFN-BR	BPRBFN-RP	BPRBFN-GDA	BPRBFN-GDM
Epochs	100	100	100	100	100	100
Goal	0	0	0	0	0	0
Max-fail	5	5	5	5	5	5
Show	25	25	25	25	25	25
Time	Inf	Inf	Inf	Inf	Inf	Inf
Scale-tol	–	20	–	–	–	–
Alpha	–	0.001	–	–	–	–
Beta	–	0.1000	–	–	–	–
Gamma	–	0.1000	–	–	–	–
Delta	–	0.0100	–	–	–	–
Min-grad	–	1.0000e-006	–	1.0000e-006	1.0000e-006	1.0000e-010
searchFcn	–	'srchcha'	–	–	–	–
Bmax	–	26	–	–	–	–
Mem-reduc	–	–	1	–	–	–
Mu	–	–	0.0050	–	–	–
Mu-dec	–	–	0.1000	–	–	–
Mu-inc	–	–	10	–	–	–
Mu-max	–	–	1.0000e+010	–	–	–
Delta-inc	–	–	–	1.2	–	–
Delta-max	–	–	–	50	–	–
Delta-dec	–	–	–	0.5	–	–
Lr-inc	–	–	–	–	1.05	–
Lr-dec	–	–	–	–	0.7	–
Mc	–	–	–	–	–	0.9
Max-perf-inc	–	–	–	–	1.0400	–

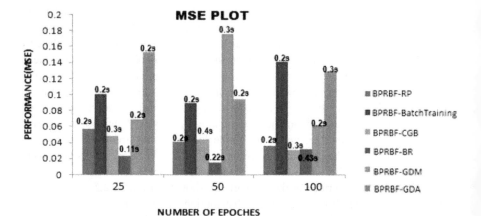

GRAPH 12.1 Mean-square-error comparison of training algorithms on BPRBFN.

TABLE 12.4
Performance Comparison of Training Algorithms

Training Function	Number of Iterations	Best Performance Time Taken (Secs)	At Epoch No.	Best Fit
BPRBFN-RP	25	0.058	24	0.88
	50	0.049	50	0.90
	100	0.04	99	0.97
BPRBFN-B	25	0.1	23	0.80
	50	0.099	50	0.82
	100	0.14	100	0.89
BPRBFN-CGB	25	0.055	25	0.82
	50	0.045	49	0.89
	100	0.03	100	0.92
BPRBFN-GDA	25	0.15	25	0.72
	50	0.99	48	0.75
	100	0.13	100	0.90
BPRBFN-GDM	25	0.07	25	0.55
	50	0.17	50	0.69
	100	0.06	99	0.80
BPRBFN-BR	25	0.02	25	0.71
	50	0.015	50	0.76
	100	0.025	100	0.87

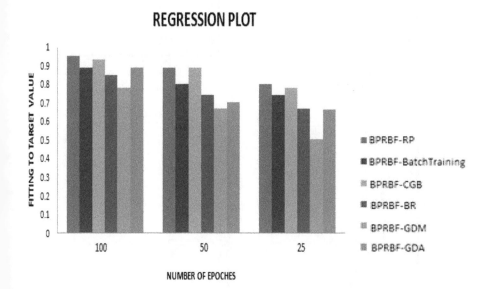

GRAPH 12.2 Cost of best fit value of training algorithms on BPRBFN.

250   Soft Computing Applications and Techniques in Healthcare

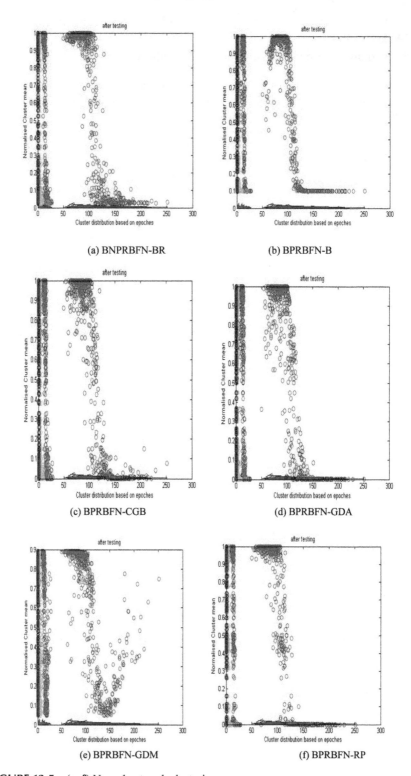

FIGURE 12.5   (a–f) Neural network clustering.

12.4.3 Discussions

Of all these, the results of trainrp were found to be compromising to a greater extent. The best target fitting value of 0.98 under 0.02 s with an error function of 0.02 was found from trainrp. The performance of trianrp for the different numbers of iterations is compared and shown in Graph 12.3 (a–c). Selecting the optimal amount of iterations is essential for the overall performance of clustering.

It is essential to note the importance of feature selection in the process of clustering. Precise conditions for clustering will lead to efficient clusters. The quality of feature selection will enhance the performance of clustering. Graph 12.3 shows the graph of R-value fit before and after clustering. Fitting is at the best when clustering is done with the selected features. Graph 12.4 indicates that the dataset has got a higher value of R, after the process of feature selection. This is because the network will have to work only in the optimised set of inputs through reduced distances after feature selection.

12.5 CONCLUSION

Analysing the cancer data to identify cancer survival ability is attempted in this work. Wisconsin breast cancer data (diagnostic) is used in the process of clustering. Clustering enables one to identify understandable patterns from the data. A modified BPRBFN-based clustering model is proposed in this work to cluster the essential features and for further analysis. Genetic algorithm-based feature selection is used to identify the 10 essential features from the 32 features of the dataset. BPRBFN network is trained using six training algorithms to analyse the network for performance under different training measures. Back propagation learning is used to redirect the error using chain rule to better performance of the network. It was analysed that, training the network using resilient back propagation gives better performance than the other algorithms. The clusters are validated using mean squared error and regression fit values.

Some implications from physicians are provided in nutshell below:

1. The proposed method can be used as a supplement to cross-check the opinions of doctors. Due to the systematic ability of the proposed algorithm, it can be used to trace back opinions, gain feedbacks and refine judgments.
2. The proposed algorithm is flexible and scalable for other medical diagnosis, provided the data is available for training and testing purposes. The algorithm adopts learning mechanism, which can improve the accuracy by mitigating the errors.
3. It can also act as an aid for patients to test their symptoms and gain knowledge of the diagnosis process.
4. The algorithm could be embedded with healthcare management system to offer a systematic diagnosis and detection of various diseases.
5. Finally, physicians must work with machine learning experts to develop such systems that could aid doctors in faster and accurate detection of diseases.

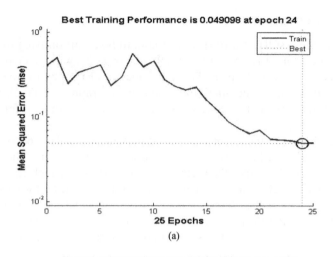

(a)

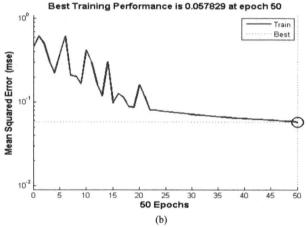

(b)

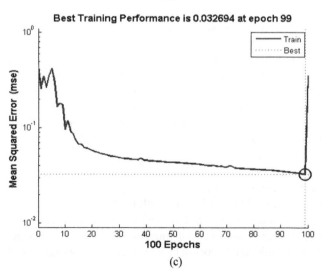

(c)

GRAPH 12.3 (a, b and c) Performance of trainrp under 25, 50, 100 epochs.

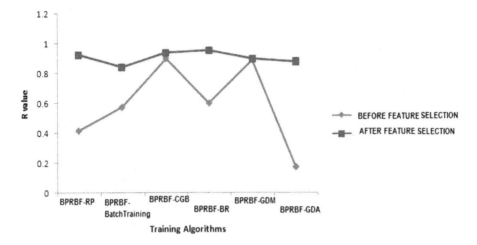

GRAPH 12.4 Effect of feature selection on the dataset.

The future of this work can be to determine the cancer patient's survivability from the cluster. Also, other optimisation techniques like ant colony optimisations and particle swarm optimisation can be used for feature selection and to optimise the group. Finally, deep learning models can be integrated with these models for better detection.

ACKNOWLEDGEMENT

The authors thank the funding agencies for financial aid from grant nos. (ERIP/ER/ 1203080/M/01/1569; 09/1095(0026)18-EMR-I, F./2015-17/RGNF-2015-17-TAM-83 and SR/FST/ETI-349/2013).

REFERENCES

1. Tjandra, J., & Collins, J. P., Breast surgery. In G. J. A. Clunie, J. Tjandra, J. A. Smith, & A. H. Kaye (eds.), *Textbook of Surgery*, 3rd ed. Singapore: Blackwell, 2008.
2. Lippman, M. E., Dickson, R. B., Bates, S., Knabbe, C., Huff, K., Swain, S., McManaway, M., Bronzert, D., Kasid, A., & Gelmann, E. P. (1986). Autocrine and paracrine growth regulation of human breast cancer. *Breast Cancer Research and Treatment*, 7(2), 59–70.
3. Dua, D., & Graff, C. *UCI Machine Learning Repository*, Irvine, CA: University of California, School of Information and Computer Science, 2019. http://archive.ics.uci.edu/ml.
4. John, G. H., Kohavi R., & Fleger, K. P. Irrelevant feature and the subset selection problem. In Proceedings of the 11th International Conference Machine Learning, pp. 121–129, 1994.
5. Lee, In-Hee, Lushington, G. H., & Visvanathan, M. (2011). A filter-based feature selection approach for identifying potential biomarkers for lung cancer. *Journal of Clinical Bioinformatics*, 1(11): 11.
6. Kong, J., Wang, S., & Wahba, G. (Sep 2014). Using distance covariance for improved variable selection with application to genetic risk models, 14077297v3. *Statistics in Medicine*, 34(10): 1708–1720.

7. Ramírez, J., Górriz, J. M., Salas-Gonzalez, D., Romero, A., López, M., Álvarez, I., & Gómez-Río, M. (2013). Computer-aided diagnosis of Alzheimer's type dementia combining support vector machines and discriminant set of features. Elsevier, *Information Sciences, 237*, 59–72.

8. Chen, H.-L., Yang, B., Liu, J., & Liu, D-Y. (2011). A support vector machine classifier with rough set-based feature selection for breast cancer diagnosis. Elsevier, *Expert Systems with Applications, 38*, 9014–9022.

9. Karegowda, A., Jayaram, M. A., & Manjunath, A. S. (2010). Feature subset selection problem using wrapper approach in supervised learning. *International Journal of Computer Applications, 1*(7), 0975–8887.

10. Hsu, H.-H., Hsieh, C-W., & Lu, M-D. (2011). Hybrid feature selection by combining filters and wrappers. Elsevier, *Expert Systems with Applications, 38*, 8144–8150.

11. Fana, C-Y., Changb, P-C., Linb, J-J., & Hsiehb, J. C. (2011). A hybrid model combining case-based reasoning and fuzzy decision tree for medical data classification. Elsevier, *Applied Soft Computing, 11*, 632–644.

12. Unler, A., Murat, A., & Chinnam, R. B. (2011). Mr2pso: a maximum relevance minimum redundancy feature selection method based on swarm intelligence for support vector machine classification. Elsevier, *Information Sciences, 181*, 4625–4641.

13. Piao, Y., & Piao, M.. (2012). Kiejung Park and Keun Ho Ryu, an ensemble correlation-based gene selection algorithm for cancer classification with gene expression data. *Bioinformatics, 28*(24), 3306–3315.

14. Dubey, A. K., Gupta, U., & Jain, S. (2016). Analysis of k-means clustering approach on the breast cancer isconsin dataset. *International Journal of Computer-assisted Radiology and Surgery, 11*(11), 2033–2047.

15. Shukla, N., Hagenbuchner, M., Win, K. T., & Yang, J. (2018). Breast cancer data analysis for survivability studies and prediction. *Computer Methods and Programs in Biomedicine, 155*, 199–208.

16. Montazeri, M., Montazeri, M., Montazeri, M., & Beigzadeh, A. (2016). Machine learning models in breast cancer survival prediction. *Technology and Health Care, 24*(1), 31–42.

17. Kumar, V., Mishra, B. K., Mazzara, M., & Verma, A. (2019). Prediction of Malignant & Benign Breast Cancer: A Data Mining Approach in Healthcare Applications. arXiv preprint arXiv:1902.03825.

18. Pritom, A. I., Munshi, M. A. R., Sabab, S. A., & Shihab, S. Predicting breast cancer recurrence using effective classification and feature selection technique. In 19th International Conference on Computer and Information Technology (ICCIT), pp. 310–314. IEEE, December 2016

19. Wuniri, Q., Huangfu, W., Liu, Y., Lin, X., Liu, L., & Yu, Z. (2019). A generic-driven wrapper embedded with feature-type-aware hybrid Bayesian classifier for breast cancer classification. *IEEE Access, 7*, 119931–119942.

20. Ahmad, F., Isa, N. A. M., Hussain, Z., Osman, M. K., & Sulaiman, S. N. (2015). A GA-based feature selection and parameter optimization of an ANN in diagnosing breast cancer. *Pattern Analysis and Applications, 18*(4), 861–870.

21. Zeynu, S., & Patil, S. (2018). Prediction of chronic kidney disease using data mining feature selection and ensemble method. *International Journal of Data Mining in Genomics & Proteomics, 9*(1), 1–9.

22. Aličković, E., & Subasi, A. (2017). Breast cancer diagnosis using GA feature selection and rotation forest. *Neural Computing and Applications, 28*(4), 753–763.

23. Akay, M. F. (2009). Support vector machines combined with feature selection for breast cancer diagnosis, *Expert Systems with Applications, 36*(2), 3240–3247.

Cluster Analysis of Breast Cancer Data Using Modified BP-RBFN 255

24. Liu, Y.-Q., Wang, C., & Zhang, L. Decision tree based predictive models for breast cancer survivability on imbalanced data. In Proceedings of the 2009 3rd International Conference on Bioinformatics and Biomedical Engineering, pp. 1–4, Beijing, China, June 2009.
25. Sheikhpour, R., Sarram, M. A., & Sheikhpour, R. (2016). Particle swarm optimization for bandwidth determination and feature selection of kernel density estimation based classifiers in diagnosis of breast cancer. *Applied Soft Computing, 40*, 113–131.
26. Kouser, K., & Priyam, A. (2018). Feature selection using genetic algorithm for clustering high dimensional data. *International Journal of Engineering & Technology, 7*(2.11), 27–30.
27. Zhiwen, Yu, & Wong. Hau-San. Genetic-based K-means algorithm for selection of feature variables. In 18th International Conference on Pattern Recognition (ICPR'06), vol. 2, pp. 744–747. IEEE, 2006.
28. Jeyasingh, S., & Veluchamy, M. (2017). Modified bat algorithm for feature selection with the isconsin diagnosis breast cancer (WDBC) dataset. *Asian Pacific Journal of Cancer Prevention: APJCP, 18*(5), 1257.
29. Aalaei, S., Shahraki, H., Rowhanimanesh, A., & Eslami, S. (2016). Feature selection using genetic algorithm for breast cancer diagnosis: experiment on three different datasets. *Iranian Journal of Basic Medical Sciences, 19*(5), 476.
30. Zarbakhsh, P., & Addeh, A. (2018). Breast cancer tumor type recognition using graph feature selection technique and radial basis function neural network with optimal structure. *Journal of Cancer Research and Therapeutics, 14*(3), 625.
31. Dhahri, H., Maghayreh, E. A., Mahmood, A., Elkilani, W., & Nagi, M. F. (2019). Automated breast cancer diagnosis based on machine learning algorithms, *Journal of Healthcare Engineering*, Article ID 4253641, *2019*, 1–11. https://doi.org/10.1155/2019/4253641.
32. Wang, A., An, N., Yang, J., Chen, G., Li, L., & Alterovitz, G. (2017). Wrapper-based gene selection with Markov blanket. *Computers in Biology and Medicine, 81*, 11–23.
33. Prakash, M., & Murty, M. N. (1997). Feature selection to improve classification accuracy using a genetic algorithm. *Journal of the Indian Institute of Science*, Jan–Feb, 77, 85–93.
34. Anbarasi, M., Anupriya, E., & Iyengar, N. Ch. S. N. (2010). Enhanced prediction of heart disease with feature subset selection using genetic algorithm. *International Journal of Engineering Science and Technology, 2*(10), 5370–5376.
35. Dy, J. G., Kak, C. E. B. A., Broderick, L. S., & Aisen, A. M. (2003). Unsupervised feature selection applied to content-based retrieval of lung images, *IEEE Transactions On Pattern Analysis and Machine Intelligence, 25*(3), 373–378.
36. Pokorný, P., & Dostál, P. (2008). Cluster analysis and neural network. In Technical Computing Prague 2008, Sborník Píspvk 16. Roníku Konference (pp. 25–34).
37. Sheela, K. G., & Deepa, S. N. (2013). Review on methods to fix number of hidden neurons in neural networks. *Mathematical Problems in Engineering*, Article ID 425740, *2013*, 1–11. https://doi.org/10.1155/2013/425740.
38. Markopoulos, A. P., Georgiopoulos, S., & Manolakos, D. E. (2016). On the use of back propagation and radial basis function neural networks in surface roughness prediction. *Journal of Industrial Engineering International, 12*(3), 389–400.
39. Kizi, O., & Uncuoglu, E. (October 2005). Comparison of three back propagation algorithms for two case studies. *Indian Journal of Engineering & Materials Sciences, 12*, 434–442.
40. MATLAB® version 9.3.0. Natick, Massachusetts: The MathWorks Inc., r2017b.
41. Vacic, V. (2005). Summary of training functions in Matlab's NN toolbox. Computer Science Department at the University of California, http://www.cs.ucr.edu/~vladimir/cs171/nn_summary.pdf.

Index

Note: Locators in *italics* represent figures and **bold** indicate tables in the text.

A

Aam Aadmi Bima Yojana, 217
Acquired immune deficiency syndrome, 98
ADABoost, 4–8, *8*
Adaptive learning, 13, 246
Adaptive neuro-fuzzy inference system, 145
Administration of anaesthesia, 155–156
Agglomerative clustering, 179–180
AI in healthcare, *166*
Anaemia, 150, 159
Analog-to-digital convertor, 187
Analysis of signs or symptoms, 154
Analytic hierarchy process, 72, 74
Antireterovial treatment (ART), 98
Artificial intelligence, 164–165, 168, *169*, 182
Artificial neural network, 4, 13, 15, 25, 27
Asthma, 153, 159
Attribute ranker, 172
Auscultation, 113–114
Automatic drug discovery, 171
Ayushman Bharat, 216–218, 224

B

Bayesian networks, 147
Benedixon, 109
Breast cancer data, 236–237, **238**

C

Cancer model, 59, 61, 63
Caputo, 59
Cardiovascular disease, 113–114
Central Government Health Scheme, 217
Challenges to Healthcare, 142
Chang's Extent Analysis of FAHP, 76
Child Survival and Safe Motherhood, 218
Chronic kidney disease, 145, 237
Clustering, 148–149, 155, 177
Communication sharing information, 166–167
Competitive learning, 14, 32
Conjugate algorithms, 247
Coordinate descent, 51
Coronary bypass surgery, 153

D

Data Acquisition and Collection, 41
Data collection and processing, 1, 4
Data mining, 147–148, 183, 236
Deep learning, 13, 32–33, 253
Density estimation, 175, 237
Density-based clustering, 179, 181, *181*
Determination of drug dose, 140, 156
Diabetes, 140, 152, 159, 226
Diagnosis of diseases, 143, 149
Diagnostic, **131**, 178, 186, 192, 228
Digital spiral basics, 36–37
Drug discovery and manufacturing, 14
Dynamic spiral test, 38, 41, 45, **131**

E

Electronic medical record, 186, 189, 204, *207*
Electronic Vaccine Intelligence
  Network, 219
Emergence of machine learning, 168
Employment State Insurance Scheme, 217
Ensemble model, 50–51, 53–54
Evolutionary computing, 114
Expanded Programme of Immunization, 218
Experimental setup, 38–39, 52

F

Factor analysis, 172–173
Feasibility, 215, 231
Feature evaluator, 176–177
Feature engineering, 37, 41–42, 45, 52, 55
Feature ranker algorithm, 177
Feature selection in medicine, 171
Fixed-point theorem, 62
Forward-backwards propagation, 144
Fuzzy analytic hierarchy process, 72, 75
Fuzzy cognitive maps, 144
Fuzzy logic systems, 140–141, 145, 154
Fuzzy set theory, 73, 141, 153, 155
Fuzzy system in healthcare, 142

G

Genetic algorithm, 141–142, 146, *146*
Gradient descent training, 245

257

Index

H

Health social security, 218
Healthcare industry, 164–166, 170
Heart sound classification, 113–114, **115–116, 118, 120, 122–123**, *123*, 127
Hierarchical clustering, 179, **133**, *180*
HIV, 151, 159
Hopfield networks, 14
Hospital selection, 80
Hybrid neuro-fuzzy systems, 145

I

I-EMR, 186, 212
Improving service quality, 156
Inactivated polio vaccine, 219
Information retrieval, 148, 158

J

Janashree Bima Yojana, 217

K

K-Nearest Neighbours, 3–4, 181
Kernel Mittag Leffler, 66
Kinematic features, 37–38, 40, 42, 45–46, 52
KNN Classifiers, **120**

L

Linear regression, 1, 3, 6, 173–174, *174*
Linear DR Algorithm, 177
Linear SVC, 50–53, **53–54**
Logistic regression, 1, 50–54, **53, 119, 128**
Lung cancer, 145, 148, 159

M

Machine learning, 175–177, 179, 181, 183, 236–238, 251
Malaria, 140, 151, 230, **222**
Mean square error, 38, *248*
Medical databases, 147, 158, 171
Medical decision support systems, 145
Medical image processing, 140, 155, 212
Medical imaging, 168, 179, 183, 187, 189, 194–195, 199, 212
Medication, 156, 165–166, *166*
Median imputation, 172–173
Mobility, 150, 167
Model selection and parameter tuning, 50–52
Monitoring, 178–179, 205, 210, *211*
MSE Index, 236
Multi-criteria decision-making, 74
Multiplier effects, 215

N

Naive Bayes algorithm, 7
National Digital Health Blueprint, 224
National Digital Health Eco-system, 224
National Health Mission, 218
National Immunization Schedule, 219
National Intelligence Grid, 223
Neural gas, 11–13, 15–17, 19, 21–23, 25, 27, 29, 31–33
Neural gas approach, 16
Neural networks, 13–14, 127, 141–142, 145, 149, 165, 182, *183*, 239
Neuro-fuzzy system, 145, 151, 159
Nonlinear DR algorithm, 177
Nonparametric method, 175–176

O

Ordinary least squares, 173

P

Parametric method, 175–176
Parkinson's disease, 35, 37, 39, 40–41, 45, 47, 51, 53, 55, 145
Partition methods, 179–180, *181*
Patient diagnosis, 13, 166–167
Performed work, 41
Predicting and managing epidemic outbreaks, 14
Predictive, 4, 7, 52, **126**, 144–145, 229
Periodontal dental disease, 149, 159
Principal component analysis, **125, 132**, 172

Q

Querying in medicine, 174, *175*

R

R-value, 239, 247, 251
Radial basis function network, 238, 244
Radial velocity, 37, 45, **46**
Random forest classifier, 51–52, **54**
Real-time operation, 13
Regression analysis, 4, 173, 247
Reinforcement learning, 168, 170
Rough set theory, 148, 158

S

Screening, 149, 171, 178, 183
Selection criteria of hospital, 80
Self-organisation, 13
Self-organizing map, 14
Soft computing techniques with fuzzy systems, 143

Index

Stability test on certain point, 41
Stages involved in drug discovery, 171
Static spiral test, 38, 41, 45, **46**
Supervised learning, 4, 14, 168
Support vector machine, 4, 120, 237, **120–122,
   128–129, 131**, 134
Surgical procedure, 166–167, *168*
Sustainable development goals, 215

T

Telemedicine, 166–167, 224, 167
Text organisation, 37
Texture classification, 182
Threshold rate, 97–101, 103, 105, 107, 109–111
TOPSIS method, 71–73, 75, 77–81, 83, 85–87,
   89, 91, 93

Triangular fuzzy number, 73, **76**, 157
Triggered neuron, 15

U

Universal health coverage, 216, 218, 224
Universal Health Database, 215–217, 219,
   223–225, 227–232, *231*
Universal Health Insurance Scheme, 217
Universalisation of vaccination, 218
Unsupervised learning, 14, 168, 170, 236

W

Widrow-Hoff algorithm, 245
World Health Organization (WHO), 113,
   218, 235